THIS BOOK BELONGS TO

....

Copyright © 2022 by HarperCollinsPublishers Limited

All rights reserved. Published in the United States by Bright Matter Books, an imprint of Random House Children's Books, a division of Penguin Random House LLC, New York. Originally published in the United Kingdom by Red Shed, part of Farshore, an imprint of HarperCollinsPublishers, London, in 2022.

Bright Matter Books and colophon are registered trademarks of Penguin Random House LLC.

Visit us on the Web! rhcbooks.com

Educators and librarians, for a variety of teaching tools, visit us at RHTeachersLibrarians.com

Library of Congress Cataloging-in-Publication Data is available upon request.
ISBN 978-0-593-90333-9 (trade) — ISBN 978-0-593-90334-6 (ebook)

Written by Miranda Smith
Interior illustrations by Jenny Wren, Juan Calle, Xuan Le, Max Rambaldi, and Olga Baumert
Front cover illustration by Juan Calle
Consultancy by Professor Mike Benton
Interior design by Duck Egg Blue Limited

MANUFACTURED IN MALAYSIA
10 9 8 7 6 5 4 3 2 1

First U.S. Edition

Random House Children's Books supports the First Amendment and celebrates the right to read.

Penguin Random House LLC supports copyright. Copyright fuels creativity, encourages diverse voices, promotes free speech, and creates a vibrant culture. Thank you for buying an authorized edition of this book and for complying with copyright laws by not reproducing, scanning, or distributing any part in any form without permission. You are supporting writers and allowing Penguin Random House to publish books for every reader.

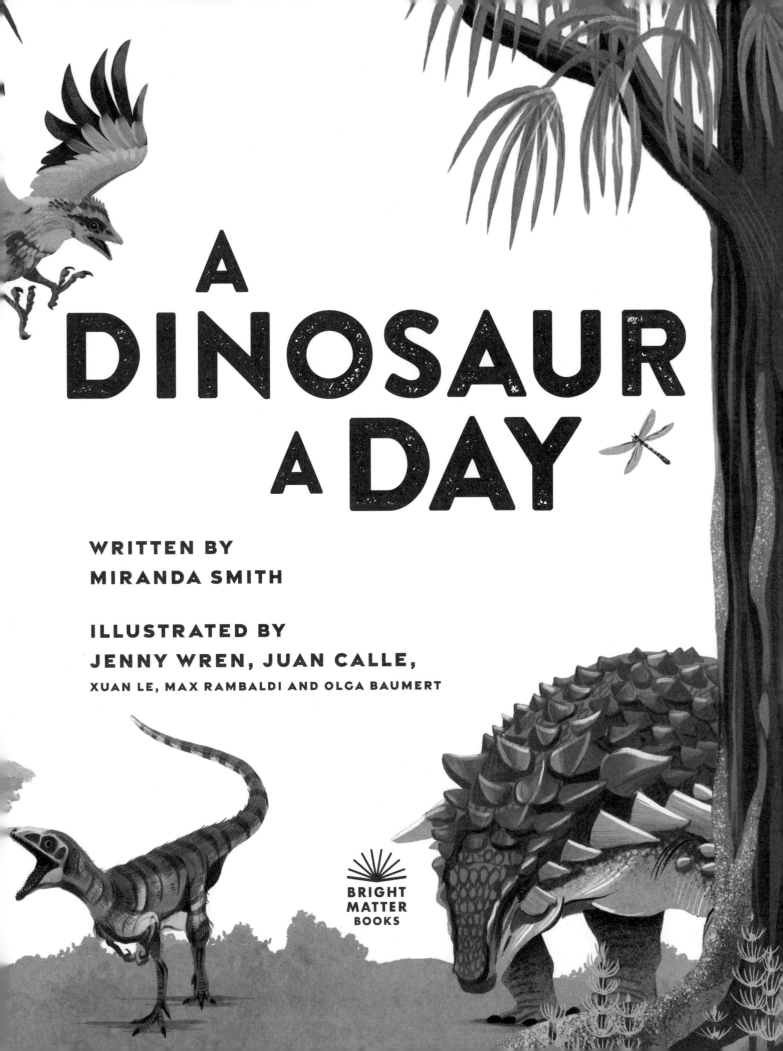

A DINOSAUR A DAY

WRITTEN BY
MIRANDA SMITH

ILLUSTRATED BY
JENNY WREN, JUAN CALLE,
XUAN LE, MAX RAMBALDI AND OLGA BAUMERT

BRIGHT MATTER BOOKS

• CONTENTS •

A world of dinosaurs	10
The rise of the dinosaurs	12

January	**14**		**July**	**110**
Some of the smallest	20		Meat-eating dinosaurs	116
Bird mimics	28		Sea reptiles	122
February	**30**		**August**	**126**
Speedy dinosaurs	36		Feathered dinosaurs	132
Insect eaters	42		Small herders	140
March	**46**		**September**	**142**
Duck-billed dinosaurs	50		Roof lizards	146
Dinosaur teeth	56		Early birds	154
April	**62**		**October**	**158**
Dinosaur weaponry	68		Bone-headed dinosaurs	164
Fish-eating dinosaurs	76		Tyrannosaur tyrants	174
May	**78**		**November**	**176**
Flying reptiles	84		Peaceful plant eaters	180
Horn-faced dinosaurs	90		Giant sauropods	186
June	**94**		**December**	**192**
Largest of them all	98		Biped hunters	196
Dinosaurs with beaks	104			

The end of the dinosaurs	210
Pronunciation guide	212
Glossary	216
Index	218

A WORLD OF DINOSAURS

For more than 160 million years, dinosaurs of all sizes roamed the planet, alongside some magnificent swimming and flying reptiles. This is your dinosaur calendar, with an awesome prehistoric creature for every day of the year. Discover a dinosaur a day for yourself – which amazing reptile is on your birthday? Or share each one with your family, friends, carers and teachers, before exploring the rest of the book to find out more about the animals that lived in this dangerous and exciting world.

THE RISE OF THE DINOSAURS

Life on Earth began at least 3.8 billion years ago. It is not known exactly how it started, but many experts believe it was as tiny organisms called microbes in the sea, and simple kinds of animals developed there from around 800 million years ago (mya). A cold climate 360 mya is believed to have caused a mass extinction, or killing off, of 70 percent of marine animals. When the world warmed up again, this triggered the appearance of the first reptiles on land and the first animals that could fly in the skies.

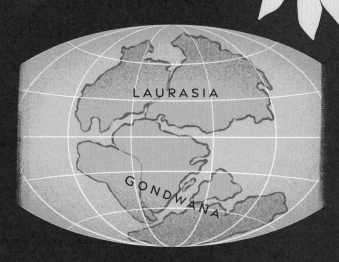

During the Triassic, 252-201 mya, a single giant landmass, Pangaea, began to break apart, forming Gondwana in the south and Laurasia in the north. On land, there were giant forests full of conifers and vines as well as dry deserts and fern-covered prairies. The first dinosaurs were small bipeds, hunting for small animals and insects in the undergrowth.

Trilobites first appeared around 520 mya in the sea.

Dinosaurs first appeared around 240 mya in the Triassic period, and reigned supreme among all other animals on Earth during the Jurassic and Cretaceous periods.

Daemonosaurus, an early dinosaur

HOW WE KNOW ABOUT THEM

Fossils are the remains of animals and plants that lived millions of years ago. Scientists called paleontologists study life in the distant past. They are able to tell us about dinosaurs because of fossils that have been found in rocks.

Sometimes dinosaur bones were covered quickly by sediment that, over time, hardened into rock. When the rock is broken open, the shape of the bones is revealed. Sometimes the forms of muscles, scales or feathers are also discovered. Trace fossils, such as burrows and tracks, show dinosaur movements or activity.

From these fossils – even the smallest part of a bone – paleontologists can tell us about how big a dinosaur was, what it ate and how it died.

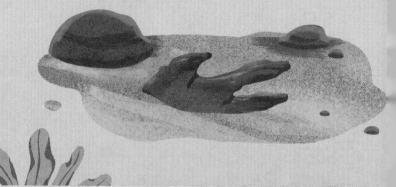

In the Jurassic, 201–145 mya, the two landmasses, Laurasia and Gondwana, moved further away from each other. Laurasia drifted north and Gondwana south. It was warmer at this time than today, with more rainfall so plants flourished. Plant-eating dinosaurs, some of the largest animals ever to exist, dominated the landscape.

During the Cretaceous, 145–66 mya, the two landmasses split up again into most of the continents we recognize today. Sea levels were higher, so there were some inland seas, such as the Western Interior Seaway in North America. Insects and flowering plants were in abundance, the number of types of mammal increased and the first birds appeared. There were more dinosaurs than ever before, developing differently depending on which continent they lived.

The sauropod Apatosaurus

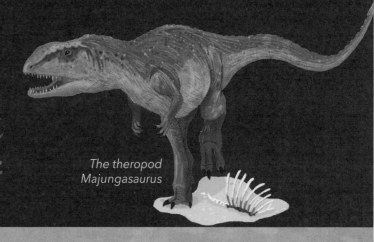

The theropod Majungasaurus

LIVING WITH DINOSAURS

Although this was the age of reptiles, not all reptiles were dinosaurs. Swimming reptiles such as Ichthyosaurus and Mosasaurus swam in the sea, while flying reptiles such as Pteranodon and Quetzalcoatlus soared through the air. During the Cretaceous, however, there were some dinosaurs, such as Spinosaurus, that could swim, and a few, such as Microraptor, that could glide through the air.

Other animals shared this busy world and needed to move fast to get out of danger. On land, there was plenty of small prey for dinosaurs and their young to hunt, including scuttling lizards, mammals, beetles and insects. Some dinosaurs searched in shallow seas or in rivers for amphibians and fish. And, everywhere, larger dinosaurs preyed on smaller dinosaurs.

Ichthyosaurus, a swimming reptile

JANUARY

• January 1st •
EORAPTOR

The earliest dinosaurs were sometimes quite small, and Eoraptor was one of the earliest and smallest. This meat eater was lightly built with hollow bones. It had fused bones in its hip, giving it the strength to run on two legs at some speed after prey. Its name means "dawn thief."

PERIOD	Late Triassic
FAMILY	early theropod
DIET	carnivore
LENGTH	3 ft.
WEIGHT	22 lb.
FINDS	Argentina

• January 2nd •
WEEWARRASAURUS

In 2018, miners at the Wee Warra opal mine in Lightning Ridge, Australia, found a unique fossil. They discovered a dinosaur jawbone that had fossilized into an opal, a multicolored precious stone. It came from a plant eater called Weewarrasaurus, a dog-sized biped that moved in herds across the ancient Cretaceous floodplains.

PERIOD	Early Cretaceous
FAMILY	Ornithopodae
DIET	herbivore
LENGTH	under 7 ft.
WEIGHT	44 lb.
FINDS	Australia

• January 3rd •
TROODON

A quick-moving hunter, Troodon's big eyes would have helped it spot small mammals, frogs and lizards in the undergrowth or even after dark. Its arms folded back like a bird's and it walked and ran on two long legs. There was a claw on each of its second toes that swiveled to trap prey.

PERIOD	Late Cretaceous
FAMILY	Troodontidae
DIET	carnivore
LENGTH	7 ft.
WEIGHT	110 lb.
FINDS	North America

• January 4th •

NIGERSAURUS

This long-necked dinosaur lived in what is now the Sahara desert, but at the time was filled with river systems and plenty of green vegetation for it to eat. It had a relatively small skull with an unusually wide mouth that contained a battery of more than 500 replaceable teeth.

PERIOD	Early Cretaceous
FAMILY	Rebbachisauridae
DIET	herbivore
LENGTH	30 ft.
WEIGHT	4.4 tons
FINDS	Africa

• January 5th •

EUSTREPTOSPONDYLUS

With its big head, large serrated teeth and short arms, this fierce meat eater hunted and scavenged on the seashore. In the middle Jurassic, Europe was a series of scattered islands and this dinosaur may well have been able to swim short distances from one to the other. It fed on smaller dinosaurs and pterosaurs, as well as marine reptiles and other sea creatures.

PERIOD	Middle Jurassic
FAMILY	Megalosauridae
DIET	carnivore
LENGTH	23 ft.
WEIGHT	1,100 lb.
FINDS	Europe

• January 6th •
BARAPASAURUS

One of the oldest sauropods discovered so far, its name means "big-legged lizard." This describes it perfectly – its thigh bone was as long as a giraffe's neck! It used its spoon-shaped teeth with serrated edges to tear leaves from tall treetops.

PERIOD	Early Jurassic
FAMILY	Cetiosauridae
DIET	herbivore
LENGTH	49 ft.
WEIGHT	15.4 tons
FINDS	Asia

• January 7th •
SHUNOSAURUS

Traveling in large herds, this slow-moving plant eater had a club with two pairs of spikes on the end of its tail. This would have packed a punch if a predator such as Gasosaurus (see p.137) got too close.

PERIOD	Middle Jurassic
FAMILY	early sauropod
DIET	herbivore
LENGTH	33 ft.
WEIGHT	1.1 tons
FINDS	Asia

• January 8th •
APATOSAURUS

This dinosaur held its whip-like tail off the ground to balance as it fed on low-lying plants. Like other sauropods, Apatosaurus swallowed gastroliths (small stones) to help break up the tough plant material in its stomach.

PERIOD	Late Jurassic
FAMILY	Diplodocidae
DIET	herbivore
LENGTH	75 ft.
WEIGHT	45.2 tons
FINDS	North America

• January 9th •

CERATOSAURUS

Although it shared its habitat with larger predators such as Allosaurus (see p.117), this dinosaur must have been a fearsome sight. It had a large horn on its nose and rows of long, curved teeth to sink into plant-eating dinosaurs. With its mouth closed, the upper teeth would have extended below its lower jaw. This and its broad, flexible tail have led to the idea that it was crocodile-like and may have been able to swim.

PERIOD	Late Jurassic
FAMILY	Ceratosauridae
DIET	carnivore
LENGTH	20 ft.
WEIGHT	1,650 lb.
FINDS	North America, Africa

SOME OF THE SMALLEST

This was a dangerous world for smaller animals. Large meat-eating dinosaurs were on the search for food all the time. Small dinosaurs only survived by being able to hide in undergrowth or up trees or, mostly, by being very quick on their feet.

• January 10th •

MICRORAPTOR

This crow-sized dinosaur was one of the smallest, and probably used the feathers on its four limbs to glide from tree to tree. It had sharp, pointed teeth to feed on the small animals and insects it caught.

PERIOD	Early Cretaceous
FAMILY	Dromaeosauridae
DIET	carnivore
LENGTH	2 ft.
WEIGHT	2 lb.
FINDS	China

• January 11th •

MOROS

A mini member of the tyrannosaur family, Moros had good hearing and eyesight. It was very fast, able to run down prey and escape easily from predators.

PERIOD	Late Cretaceous
FAMILY	Tyrannosauridae
DIET	carnivore
LENGTH	4 ft.
WEIGHT	175 lb.
FINDS	North America

• January 12th •

COMPSOGNATHUS

Lightweight and very agile, this bird-like dinosaur ran on long back legs with three-toed feet. Its large eyes helped it survive by enabling it to spot the movement of prey and predator alike.

PERIOD	Late Jurassic
FAMILY	Compsognathidae
DIET	carnivore
LENGTH	5 ft.
WEIGHT	7 lb.
FINDS	Europe

• January 13th •
SALTOPUS

The long jaws with dozens of sharp teeth were great for snatching insects out of the air. To vary its diet, Saltopus – "hopping foot" – also hunted lizards, beetles and scorpions.

PERIOD	Late Triassic
FAMILY	early theropod
DIET	carnivore
LENGTH	3 ft.
WEIGHT	2 lb.
FINDS	Europe

• January 14th •
LESOTHOSAURUS

Fast and agile, this dinosaur had large eyes and big jaw muscles to bite down on small animals and soft plants. It could sprint quickly away if needed.

PERIOD	Early Jurassic
FAMILY	Lesothosauridae
DIET	omnivore
LENGTH	7 ft.
WEIGHT	22 lb.
FINDS	Africa

• January 15th •
WANNANOSAURUS

Living in large herds provided some protection for smaller dinosaurs. Like other members of its family, this one also had a hard, flat-topped skull that it could use to head-butt a predator.

PERIOD	Late Cretaceous
FAMILY	Pachycephalosauridae
DIET	herbivore
LENGTH	less than 3 ft.
WEIGHT	10 lb.
FINDS	Asia

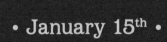

• January 16th •

STEGOSAURUS

This armored plant eater was the largest member of its family, but its small head only housed a tiny brain about the size of a lemon. Its snout was held close to the ground to feed on low-lying vegetation such as ferns and horsetails. When threatened, its impressive spiked tail could have been wielded to discourage or injure a predator such as Allosaurus (see p.117).

PERIOD	Late Jurassic
FAMILY	Stegosauridae
DIET	herbivore
LENGTH	30 ft.
WEIGHT	3.3 tons
FINDS	North America, Europe

• January 17th •

BRACHIOSAURUS

While groups of Stegosauruses browsed on low-lying plants, the peaceful Brachiosaurus was able to reach high into the trees for tender leaves. It too had a small head, but it was perched on a neck up to 52 feet long. This was one of the heaviest dinosaurs, with front legs much longer than its hind ones.

PERIOD	Late Jurassic
FAMILY	Brachiosauridae
DIET	herbivore
LENGTH	75 ft.
WEIGHT	79.4 tons
FINDS	North America, Europe, Africa

• January 18th •

PLATEOSAURUS

Moving on all fours in a herd, this plant-eating dinosaur could rear up to reach the tasty leaves at the top of tall trees and grind them with its leaf-shaped teeth. It had five-fingered hands with a large thumb claw, which it used to dig up roots to eat and also to protect itself if threatened.

PERIOD	Late Triassic
FAMILY	Plateosauridae
DIET	herbivore
LENGTH	23 ft.
WEIGHT	1,985 lb.
FINDS	Europe, North America

• January 19th •

PROCERATOSAURUS

The display crest on its snout set this hunter apart. Possibly the earliest tyrannosaur, it had a powerful jaw full of sharp, slicing teeth. Its unusually enlarged nostrils would have helped it sniff out prey as it moved on two legs through its territory.

PERIOD	Middle Jurassic
FAMILY	Proceratosauridae
DIET	carnivore
LENGTH	10 ft.
WEIGHT	220 lb.
FINDS	Europe

• January 20th •

SILVISAURUS

This plant eater's name means "woodland lizard" because it lived in shady forests. It had studded body armor and spikes for protection as it browsed on shrubs with its small, pointed teeth.

PERIOD	Middle Cretaceous
FAMILY	Nodosauridae
DIET	herbivore
LENGTH	13 ft.
WEIGHT	1.1 tons
FINDS	North America

• January 21st •

ERLIKOSAURUS

Although this dinosaur was a theropod, it belonged to a family that ate plants. With an upper jaw ending in a toothless beak, it also had elongated claws to scrape up seaweed and other plants to eat.

PERIOD	Middle Cretaceous
FAMILY	Therizinosauridae
DIET	herbivore
LENGTH	20 ft.
WEIGHT	350 lb.
FINDS	Asia

• January 22nd •

SINORNITHOIDES

The size of a turkey, this insect eater scratched for ants with its large front claws. It was a fast runner on its long legs, and used its speed to catch small animals to eat and also to get out of harm's way.

PERIOD	Early Cretaceous
FAMILY	Troodontidae
DIET	carnivore
LENGTH	4 ft.
WEIGHT	12 lb.
FINDS	Asia

• January 23rd •
MUTTABURRASAURUS

Named after a town in Australia, this herbivore munched on cycads and conifers as it moved in a herd across its territory. The bony bump on its snout may have helped its sense of smell or possibly enabled it to make warning calls. It had a toothless beak and tightly packed back teeth that formed blades to help it eat tough vegetation.

PERIOD	Early Cretaceous
FAMILY	Rhabdodontidae
DIET	herbivore
LENGTH	25 ft.
WEIGHT	3.1 tons
FINDS	Australia, Antarctica

• January 24th •

GIGANTORAPTOR

Standing tall at more than 16 feet, this extraordinary creature is the largest known beaked dinosaur. Being big would have given it significant advantages, with fewer predators and more food possibilities. It was toothless but had sharp claws, probably eating or scavenging whatever it found, animals and plants alike.

PERIOD	Late Cretaceous
FAMILY	Oviraptorosauridae
DIET	omnivore
LENGTH	26 ft.
WEIGHT	1.5 tons
FINDS	Asia

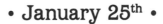

• January 25th •

IGUANODON

Large herds of Iguanodon grazed on ferns and horsetails near streams and rivers. The dinosaur walked on all fours, but could rise up on its back legs to reach tasty leaves using its sturdy tail for balance. It was the biggest in its family and had a large thumb spike that it may have used to defend itself against predatory carnivores. Remains of this successful group of animals have been found on most continents around the world.

PERIOD	Early Cretaceous
FAMILY	Iguanodontidae
DIET	herbivore
LENGTH	33 ft.
WEIGHT	5.5 tons
FINDS	Europe, Asia, Africa, North America, Australia

BIRD MIMICS

The family Ornithomimidae – the name means "bird mimics" – were ostrich-like. They were slender and lightly built, with a small head perched on top of a long neck, and had long legs and a long tail.

• January 26th •

SHENZHOUSAURUS

This early member of the family was a fast-moving predator that had teeth only in its lower jaw. Remains have been found with small stones (gastroliths) in the stomach to grind up plants.

PERIOD	Early Cretaceous
FAMILY	early Ornithomimidae
DIET	omnivore
LENGTH	5 ft.
WEIGHT	55 lb.
FINDS	Asia

• January 27th •

HARPYMIMUS

Slim, three-fingered hands and 22 teeth in the front of its lower jaw allowed Harpymimus to grip plants, insects or small animals before swallowing them. It was small and nippy when needed.

PERIOD	Early Cretaceous
FAMILY	early Ornithomimidae
DIET	omnivore
LENGTH	7 ft.
WEIGHT	275 lb.
FINDS	Asia

• January 28th •

ORNITHOMIMUS

This was an efficient hunter with good eyesight. Its beak-like jaw was toothless, and it swallowed its prey, such as insects and small animals, whole.

PERIOD	Late Cretaceous
FAMILY	Ornithomimidae
DIET	omnivore
LENGTH	13 ft.
WEIGHT	375 lb.
FINDS	North America

• January 29th •
ANSERIMIMUS

Using its large claws at the end of powerful arms, Anserimimus would have dug into the earth and termite mounds for its food. Its name means "goose mimic."

PERIOD	Late Cretaceous
FAMILY	Ornithomimidae
DIET	omnivore
LENGTH	10 ft.
WEIGHT	110 lb.
FINDS	Asia

• January 30th •
GALLIMIMUS

Traveling at speed across open, arid plains, this lightweight theropod was able to outpace predators. It hunted lizards, snakes and mammals, and also used its hands to dig for insects.

PERIOD	Late Cretaceous
FAMILY	Ornithomimidae
DIET	omnivore
LENGTH	20 ft.
WEIGHT	880 lb.
FINDS	Asia

• January 31st •
STRUTHIOMIMUS

This dinosaur's name actually means "ostrich mimic." It had longer hands than other members of the family, so it may have relied on them to grab insects, reptiles or plants. It could run up to 43 mph.

PERIOD	Late Cretaceous
FAMILY	Ornithomimidae
DIET	omnivore
LENGTH	15 ft.
WEIGHT	330 lb.
FINDS	North America

FEBRUARY

• February 1st •

XENOCERATOPS

With two long horns over its eyes and huge spikes on its large frill, it is no wonder that this dinosaur's name means "alien horned face." The frill protected it against attacks from hungry carnivores. This peaceful plant eater traveled in a herd, feeding on the tough leaves it pulled off with its beak, and ground down with many rows of cheek teeth.

PERIOD	Late Cretaceous
FAMILY	Ceratopsidae
DIET	herbivore
LENGTH	20 ft.
WEIGHT	2.2 tons
FINDS	North America

• February 2nd •

NYASASAURUS

Some experts say that this dog-sized dinosaur is the oldest true dinosaur found so far. It had an extremely long tail that measured more than half its body length and it probably moved about on two legs, reaching its small head into the undergrowth to find prey.

PERIOD	Middle Triassic
FAMILY	early theropod
DIET	carnivore
LENGTH	10 ft.
WEIGHT	130 lb.
FINDS	Africa

• February 3rd •

OURANOSAURUS

The impressive sail on the back of Ouranosaurus was formed by long spines covered in skin. It was probably used as a display either to attract mates or put off rivals. This dinosaur walked on all fours when searching for the leaves, fruit and seeds that it ate, but could run on two legs to escape predators.

PERIOD	Early Cretaceous
FAMILY	Iguanodontidae
DIET	herbivore
LENGTH	23 ft.
WEIGHT	4.4 tons
FINDS	Africa

• February 4th •

MIRAGAIA

A long neck allowed Miragaia to reach high up into the trees. This was unusual because most stegosaurs fed on low-growing shrubs near the ground, so had shorter necks than Miragaia. The four long spikes on this dinosaur's tail could do real damage to a predator.

PERIOD	Late Jurassic
FAMILY	Stegosauridae
DIET	herbivore
LENGTH	20 ft.
WEIGHT	2.4 tons
FINDS	Europe

• February 5th •

PNEUMATORAPTOR

The name of this small dinobird means "air thief." It could not fly, but the many air cavities in its bones would have made it very lightweight. For all of its bird-like appearance, this was a fierce and active predator, chasing down small animals and lizards, and probably scavenging leftover kills from the larger theropods that shared its environment.

PERIOD	Late Cretaceous
FAMILY	Dromaeosauridae
DIET	carnivore
LENGTH	3 ft.
WEIGHT	22 lb.
FINDS	Europe

• February 6th •

DEINOCHEIRUS

This extraordinary hump-backed dinosaur was built for size, not speed. Its long, horse-like snout did not house teeth, but was used to stir up soft plants or fish at the bottom of streams and lakes. These were swallowed whole and ground up in its stomach by small stones called gastroliths. Its long claws were used to pull down branches or dig for food on land.

PERIOD	Late Cretaceous
FAMILY	Deinocheiridae
DIET	omnivore
LENGTH	39 ft.
WEIGHT	7.7 tons
FINDS	Asia

• February 7th •

AUSTROSAURUS

The titanosaurs include some of the largest land animals ever to have existed. This long-necked dinosaur was one of the smallest in the family, but it was able to pull down leaves and branches from the tallest conifers that flourished in the area's cool, wet climate.

PERIOD	Early Cretaceous
FAMILY	Titanosauridae
DIET	herbivore
LENGTH	49 ft.
WEIGHT	16 tons
FINDS	Australia

• February 8th •

ACROCANTHOSAURUS

This large meat eater is known to have hunted down large plant-eating dinosaurs. Its powerful back legs would have helped it chase down its prey. One of the biggest in its family, it had a big head, sharp teeth and a long slender tail for balance as it ran.

PERIOD	Early Cretaceous
FAMILY	Carcharodontosauridae
DIET	carnivore
LENGTH	38 ft.
WEIGHT	7.7 tons
FINDS	North America

SPEEDY DINOSAURS

The fastest dinosaurs were hunters on two legs moving quickly to catch fast-moving prey. But some smaller carnivores also needed a turn of speed to escape the clutches of fierce predators.

• February 9th •
CONCAVENATOR

With the long legs of a sprinter and tiny feet, Concavenator would have covered firm, dry ground fast. Its favorite prey were early mammals, but it also liked to catch small dinosaurs and crocodiles.

PERIOD	Early Cretaceous
FAMILY	Carcharodontosauridae
DIET	carnivore
LENGTH	20 ft.
WEIGHT	1.7 tons
FINDS	Europe

• February 10th •
DELTADROMEUS

This agile predator could run as fast as a greyhound. It needed to do so to catch smaller prey but also to escape large carnivores such as Spinosaurus (see p.80).

PERIOD	Late Cretaceous
FAMILY	Noasauridae
DIET	carnivore
LENGTH	26 ft.
WEIGHT	1.7 tons
FINDS	Africa

• February 11th •
SINOCALLIOPTERYX

Its powerful hind legs meant it could move swiftly after lizards and mammals. They also helped it leap to snatch flying reptiles from the air.

PERIOD	Early Cretaceous
FAMILY	Compsognathidae
DIET	carnivore
LENGTH	8 ft.
WEIGHT	44 lb.
FINDS	Asia

• February 12th •
ALBERTADROMEUS

The smallest plant eater in its habitat, this turkey-sized dinosaur needed to be quick because it was a popular snack for many of the carnivores sharing its home.

PERIOD	Late Cretaceous
FAMILY	Parksosauridae
DIET	herbivore
LENGTH	5 ft.
WEIGHT	35 lb.
FINDS	North America

• February 13th •
BAMBIRAPTOR

A quick, fierce hunter of small mammals and reptiles, this bird-like dinosaur had a killer claw on each foot and very sharp teeth.

PERIOD	Late Cretaceous
FAMILY	Dromaeosauridae
DIET	carnivore
LENGTH	3 ft.
WEIGHT	7 lb.
FINDS	North America

• February 14th •
GORGOSAURUS

Closely related to Tyrannosaurus (see p.113), this carnivore had huge jaws in its massive skull. It was lighter and, when young, able to run much faster than its famously slow-moving relative.

PERIOD	Late Cretaceous
FAMILY	Tyrannosauridae
DIET	carnivore
LENGTH	30 ft.
WEIGHT	3.3 tons
FINDS	North America

• February 15th •

BOREALOPELTA

This armored dinosaur was the size of a tank and armed with fearsome spikes. Hunted by fierce carnivores including Acrocanthosaurus (see *p.35*), its skin color helped to hide it among the forest trees. There, its reddish-brown upper body and countershaded paler underside blended in with the shadows. It fed mainly on leaves, but also ate wood and charcoal.

PERIOD	Early Cretaceous
FAMILY	Nodosauridae
DIET	herbivore
LENGTH	18 ft.
WEIGHT	2.2 tons
FINDS	North America

• February 16th •

ALWALKERIA

This was a small biped, running quickly on its back legs and using its tail for balance. Alwalkeria lived in an area around an ancient lake, so there would have been plenty of opportunity for all kinds of prey. It is thought that this very early dinosaur had a varied diet of insects, small animals and soft plants. Its front teeth were slim and straight, while the rest were curved backward, an unusual combination.

PERIOD	Late Triassic
FAMILY	early dinosaur
DIET	omnivore
LENGTH	5 ft.
WEIGHT	4 lb.
FINDS	Asia

• February 17th •

NOMINGIA

This dinosaur was not able to fly but certainly showed some bird-like features. The impressive fan on the end of Nomingia's tail would have been used to attract a mate, much like today's peacock. This was the first dinosaur to have been found with a pygostyle, which was made up of five fused tail bones to support the fan at the end. With its feathered body, long legs and clawed hands, it was an unusual sight.

PERIOD	Late Cretaceous
FAMILY	Oviraptoridae
DIET	omnivore
LENGTH	6 ft.
WEIGHT	46 lb.
FINDS	Asia

• February 18th •
ARISTOSUCHUS

Despite its name, which means "brave crocodile," this theropod was a bird-like dinosaur with hollow bones, long back legs and three-toed feet. It was a swift hunter in the woodlands of what is now western Europe. Its prey included frogs, insects, small mammals, lizards and early birds that it snapped up with a jaw full of sharp, needle-like teeth.

PERIOD	Early Cretaceous
FAMILY	Compsognathidae
DIET	carnivore
LENGTH	7 ft.
WEIGHT	66 lb.
FINDS	Europe

• February 19th •
DIABLOCERATOPS

With those two spectacularly large curved horns at the top of its frill and others around its head, this "devilish horn face" would have been scary enough to put off most predators. The distance from the tip of its beak to the back of its frill measured 3 feet! Its deep snout helped it to locate the plants it fed on, nipping off leaves with its beak as it moved along in a herd.

PERIOD	Late Cretaceous
FAMILY	Ceratopsidae
DIET	herbivore
LENGTH	18 ft.
WEIGHT	2.2 tons
FINDS	North America

INSECT EATERS

Most of these insectivores had specialized claws to dig out prey from the ground or hiding places such as fallen trees. Many were tiny themselves, and needed good eyesight and quick reactions to catch the insects.

• February 20th •
LINHENYKUS

This bug eater lived in the desert area of what is now Inner Mongolia. It used the unique single-clawed finger on each arm to dig into termite and ant nests for a tasty snack.

PERIOD	Late Cretaceous
FAMILY	Alvarezsauridae
DIET	carnivore
LENGTH	3 ft.
WEIGHT	4 lb.
FINDS	Asia

• February 21st •
MAHAKALA

A tiny biped with short arms, this dinobird was one of the earliest raptors. Like its relatives, it had a sickle claw on each foot to pin down its insect and small animal prey.

PERIOD	Late Cretaceous
FAMILY	Dromaeosauridae
DIET	carnivore
LENGTH	2 ft.
WEIGHT	2 lb.
FINDS	Asia

• February 22nd •
ALBERTONYKUS

The size of a chicken, this insectivore had short arms that were perfect for digging. It probably feasted on termites and other insects in decaying wooden logs.

PERIOD	Late Cretaceous
FAMILY	Alvarezsauridae
DIET	carnivore
LENGTH	2 ft.
WEIGHT	11 lb.
FINDS	North America

• February 23rd •
MONONYKUS

Hunting at night, Mononykus' large eyes helped it to spot prey and avoid predators. As well as insects, it would catch lizards and small mammals. Its name means "single claw."

PERIOD	Late Cretaceous
FAMILY	Alvarezsauridae
DIET	carnivore
LENGTH	3 ft.
WEIGHT	8 lb.
FINDS	Asia

• February 24th •
ALVAREZSAURUS

One finger on each hand of this dinosaur had an enlarged claw to dig for the insects that it snatched up with the small teeth at the front of its snout. It had long legs so would have been a fast runner.

PERIOD	Late Cretaceous
FAMILY	Alvarezsauridae
DIET	carnivore
LENGTH	7 ft.
WEIGHT	7 lb.
FINDS	South America

• February 25th •
JURAVENATOR

Apart from insects, this small theropod would also have eaten fish and lizards. It had large eyes so may even have hunted around shallow lagoons and coasts in the twilight or at night.

PERIOD	Late Jurassic
FAMILY	Compsognathidae
DIET	carnivore
LENGTH	2 ft.
WEIGHT	1 lb.
FINDS	Europe

• February 26th •

ANZU

This large theropod looked like an ostrich and has been nicknamed "the chicken from hell," as skeletons of three Anzus were found in the Hell Creek Formation in North America. It lived on the humid floodplains there at the time, hunting small animals that it grasped with its large, sharp claws. It also fed on soft plants.

PERIOD	Late Cretaceous
FAMILY	Caenagnathidae
DIET	omnivore
LENGTH	10 ft.
WEIGHT	495 lb.
FINDS	North America

• February 27th •

KAMUYSAURUS

Like other members of its family, Kamuysaurus wandered in a herd, grazing on all fours on low-lying bushes and other plants. It may, however, have been able to run away from threatening predators on its two back legs, using its tail as balance. It had a horn-like beak that contained hundreds of teeth for grinding up the tough vegetation it found in the arid area where it lived near the ocean.

PERIOD	Late Cretaceous
FAMILY	Hadrosauridae
DIET	herbivore
LENGTH	26 ft.
WEIGHT	6.6 tons
FINDS	Asia

• February 28th •

LEAELLYNASAURA

For a small dinosaur, this plant eater had an amazingly long tail with more than 70 vertebrae – about 75 percent of its body length. It would have been able to keep warm in cold weather by wrapping the tail right around its body. This unusual dinosaur also had a big brain, and large eyes that would have helped find its way in dim light.

PERIOD	Middle Cretaceous
FAMILY	Hypsilophodontidae
DIET	herbivore
LENGTH	7 ft.
WEIGHT	18 lb.
FINDS	Australia

• February 29th •

KOL

Possibly the largest member of a strange family of feathered theropods, this dinosaur had thumbs with huge claws that it probably used to dig out termites and other insects from decaying tree trunks. It also hunted down lizards and small mammals.

PERIOD	Late Cretaceous
FAMILY	Alvarezsauridae
DIET	carnivore
LENGTH	8 ft.
WEIGHT	44 lb.
FINDS	Asia

MARCH

• March 1st •
MICROPACHYCEPHALOSAURUS

One of the smallest dinosaurs, this biped has the longest name – a name that means "small, thick-headed lizard"! Despite a tough, bony head like other members of the family, its best defense would have been to scamper away from predators as fast as its legs would carry it.

PERIOD	Late Cretaceous
FAMILY	Pachycephalosauridae
DIET	herbivore
LENGTH	3 ft.
WEIGHT	10 lb.
FINDS	Asia

• **March 2nd** •

INCISIVOSAURUS

Often nicknamed "Bunnysaurus," this early turkey-sized oviraptor was a curious sight. Feathered and bird-like, it had two big, flat front teeth, much like those that rabbits use for snipping plant food. It was probably mostly vegetarian but may also have feasted on small animals and dinosaur eggs.

PERIOD	Early Cretaceous
FAMILY	Oviraptoridae
DIET	omnivore
LENGTH	3 ft.
WEIGHT	13 lb.
FINDS	Asia

• March 3rd •

TIANYULONG

About the size of a cat and with a long, fuzzy tail, this dinosaur moved on two legs in its search for plants and insects to eat. Its feathers were unusual because it belonged to a group that included the armored dinosaurs, which did not have feathers.

PERIOD	Early Cretaceous
FAMILY	Heterodontosauridae
DIET	omnivore
LENGTH	2 ft.
WEIGHT	9 lb.
FINDS	Asia

• March 4th •

RAPETOSAURUS

This tall titanosaur had a very long neck on its huge, elephant-like body. This was so it could reach high into the foliage for tasty leaves that it nipped off with its small, pencil-like teeth. It was one of the last of its family but was a little different from the others in appearance. Its long skull with nostrils at the top was similar to that of Diplodocus (see p.57), which belonged to a different family.

PERIOD	Late Cretaceous
FAMILY	Titanosauridae
DIET	herbivore
LENGTH	49 lb.
WEIGHT	15.4 tons
FINDS	Africa

DUCK-BILLED DINOSAURS

Hadrosaurs are so named because of their flat, long snout ending in a toothless beak. These peaceful duckbills were the sheep of the dinosaur world and most called using hollow crests to warn of danger to the herd.

• March 5th •

TSINTAOSAURUS

The long, horn-shaped crest on top of its skull has earned this dinosaur the description of "duck-faced unicorn." It was probably preyed on by tyrannosaurs.

PERIOD	Late Cretaceous
FAMILY	Hadrosauridae
DIET	herbivore
LENGTH	33 ft.
WEIGHT	3.3 tons
FINDS	Asia

• March 6th •

PARASAUROLOPHUS

Its crest swept backward from the head with hollow tubes running from the snout. It would have been able to produce loud trumpeting sounds.

PERIOD	Late Cretaceous
FAMILY	Hadrosauridae
DIET	herbivore
LENGTH	36 ft.
WEIGHT	4.4 tons
FINDS	North America

• March 7th •

SAUROLOPHUS

Like other hadrosaurs, Saurolophus snipped off leaves and twigs with its beak. Its bank of grinding back teeth then broke these down by sliding the top and bottom jaws against one another.

PERIOD	Late Cretaceous
FAMILY	Hadrosauridae
DIET	herbivore
LENGTH	39 ft.
WEIGHT	2.2 tons
FINDS	Asia, America

• March 8th •
KRITOSAURUS

Unusually, this hadrosaur did not have a crest, but it did have a bony lump on its snout. It browsed on plants at different levels, using its tail for balance.

PERIOD	Late Cretaceous
FAMILY	Hadrosauridae
DIET	herbivore
LENGTH	36 lb.
WEIGHT	3.3 tons
FINDS	North America

• March 9th •
CORYTHOSAURUS

The helmet-shaped crest on the head of this dinosaur may have been used to boom signals to attract a female or to warn the herd about the approach of a hungry predator.

PERIOD	Late Cretaceous
FAMILY	Hadrosauridae
DIET	herbivore
LENGTH	33 ft.
WEIGHT	6.6 tons
FINDS	North America

• March 10th •
LAMBEOSAURUS

This hadrosaur had a unique, hatchet-shaped crest. Its narrow snout ended in a wide, blunt beak, and the males may have had larger crests than the females.

PERIOD	Late Cretaceous
FAMILY	Hadrosauridae
DIET	herbivore
LENGTH	31 ft.
WEIGHT	3.3 tons
FINDS	North America

• March 11th •
LAJASVENATOR

Related to Carcharodontosaurus (see p.68) and some of the other giant flesh eaters, this was one of the smallest in its very fierce family and the earliest known to have lived in what is now South America. It would have hunted animals and dinosaurs in a land full of tropical forests and lakes teeming with fish, crocodiles and turtles.

PERIOD	Early Cretaceous
FAMILY	Carcharodontosauridae
DIET	carnivore
LENGTH	16 ft.
WEIGHT	1,100 lb.
FINDS	South America

• March 12th •
BOROGOVIA

This small troodont is named after creatures called "borogoves" that appear in "Jabberwocky," a nonsense poem by Lewis Carroll. This predator was agile and swift, hunting small prey animals such as lizards and mammals. Unlike other members of its family such as Troodon (see p.16), it did not have a sickle-shaped claw on its foot.

PERIOD	Late Cretaceous
FAMILY	Troodontidae
DIET	carnivore
LENGTH	7 ft.
WEIGHT	44 lb.
FINDS	Asia

• March 13th •

LOPHOSTROPHEUS

This dinosaur is significant because it survived the Triassic–Jurassic extinction in which more than half of all living things on Earth died out. There are very few dinosaurs known from this time period. A medium-sized theropod, it was an agile and fast biped that hunted in the swamps and marshes for its animal prey.

PERIOD	Late Triassic/Early Jurassic
FAMILY	Coelophysidae
DIET	carnivore
LENGTH	16 ft.
WEIGHT	440 lb.
FINDS	Europe

• March 14th •

ORNITHOLESTES

Very light on its feet, this small, long-legged carnivore was very quick in its pursuit of the small mammals, lizards and hatchling dinosaurs it liked to eat. It probably also scavenged and its name means "bird robber." Its large eyes were set in a small head perched on a long neck and it had a very long tail that made up half its length.

PERIOD	Late Jurassic
FAMILY	Coeluridae
DIET	carnivore
LENGTH	7 ft.
WEIGHT	26 lb.
FINDS	North America

• March 15th •

DROMAEOSAURUS

About the size of a wolf and traveling in packs, this large-headed predator had a mouth full of serrated teeth and a sharp sickle claw on each hind foot to bring down prey. Its very large eyes gave it razor-sharp vision, and good hearing and smell helped it target herbivores. It ran down and leaped on prey, seizing it with its hand claws while delivering slashing kicks from a hind leg.

PERIOD	Late Cretaceous
FAMILY	Dromaeosauridae
DIET	carnivore
LENGTH	7 ft.
WEIGHT	33 lb.
FINDS	North America

• March 16th •

EDMONTOSAURUS

However quickly it ran, this peaceful hadrosaur probably could not have moved fast enough to escape a determined and hungry pack of dromaeosaurs. Weaving through the trees, an adult would have tried to protect its young from attack while inflating skin near its nose to bellow a warning to other dinosaurs in its herd.

PERIOD	Late Cretaceous
FAMILY	Hadrosauridae
DIET	herbivore
LENGTH	43 ft.
WEIGHT	4.4 tons
FINDS	North America

DINOSAUR TEETH

From sharp, serrated ones for ripping flesh to batteries for grinding tough plants, dinosaur teeth varied widely. Most were replaced when damaged. Some dinosaurs would go through thousands in a lifetime.

• March 17th •
HERRERASAURUS

This terrifying dinosaur had a sliding lower jaw that allowed it to hold on to prey more easily. Its strong jaws and long, curved teeth made short work of its large reptile prey.

PERIOD	Late Triassic
FAMILY	Herrerasauridae
DIET	carnivore
LENGTH	20 ft.
WEIGHT	770 lb.
FINDS	South America

• March 18th •
MONOLOPHOSAURUS

Many sharp, serrated teeth filled this meat eater's long crested snout. It may have hunted in a pack, chasing down sauropod prey such as Mamenchisaurus (see p. 98).

PERIOD	Middle Jurassic
FAMILY	Megalosauridae
DIET	carnivore
LENGTH	23 ft.
WEIGHT	1,540 lb.
FINDS	Asia

• March 19th •
HETERODONTOSAURUS

This extraordinary small dinosaur had three kinds of teeth – for biting, tearing and grinding! There were two tusks and small peg-like teeth at the front of the jaw, and grinding teeth toward the back.

PERIOD	Early Jurassic
FAMILY	Heterodontosauridae
DIET	herbivore
LENGTH	4 ft.
WEIGHT	22 lb.
FINDS	Africa

• March 20th •
DIPLODOCUS

For a big dinosaur, this plant eater had small teeth. Forward-pointing and easily damaged, it is estimated one tooth was replaced every 35 days of its life.

PERIOD	Late Jurassic
FAMILY	Diplodocidae
DIET	herbivore
LENGTH	89 ft.
WEIGHT	22 tons
FINDS	North America

• March 21st •
CAMARASAURUS

A keen sense of smell would have guided this herbivore to the very best places to eat high in the trees. There, it would have used its large, sharp-pointed teeth to crop tough vegetation.

PERIOD	Late Jurassic
FAMILY	Camarasauridae
DIET	herbivore
LENGTH	75 ft.
WEIGHT	22 tons
FINDS	North America, Europe

• March 22nd •
MEGALOSAURUS

This large beast's narrow skull housed long, sharp, blade-like teeth to slice through prey. It would also have scavenged, taking advantage of dead dinosaur finds.

PERIOD	Middle Jurassic
FAMILY	Megalosauridae
DIET	carnivore
LENGTH	30 ft.
WEIGHT	3.3 tons
FINDS	Europe

• March 23rd •

CENTROSAURUS

The large, forward-pointing horn on its nose and the two hooked spikes on its frill were a great defense against a predator. This herbivore ate low-lying plants as it moved across the plains and through the woodlands of what is now western North America. An entire herd, including youngsters, was found in a fossil riverbed, possibly drowned together while trying to cross the flooded waters.

PERIOD	Late Cretaceous
FAMILY	Ceratopsidae
DIET	herbivore
LENGTH	20 ft.
WEIGHT	3.3 tons
FINDS	North America

• March 24th •

RHABDODON

With a blunt head ending in a beak, this dinosaur probably walked on its strong hind legs, snipping foliage from trees and flowering shrubs as it moved with its herd. In the warm and humid atmosphere of the time, Rhabdodon would have been easy pickings for the large carnivores that shared its habitat, so there was safety in numbers.

PERIOD	Late Cretaceous
FAMILY	Rhabdodontidae
DIET	herbivore
LENGTH	15 ft.
WEIGHT	990 lb.
FINDS	Europe

• March 25th •
VELOCISAURUS

Although Velocisaurus was short, stout and the smallest member of its family, it had long legs and could run very fast. Its robust third toe indicates that it spent much of its life running so it was given a name that means "swift lizard." As well as chasing prey such as small mammals, it probably used its speed to get away from the larger theropods that hunted in its habitat.

PERIOD	Late Cretaceous
FAMILY	Noasauridae
DIET	carnivore
LENGTH	4 ft.
WEIGHT	22 lb.
FINDS	South America

• March 26th •
PROTOCERATOPS

This sheep-sized dinosaur liked to munch on cycads and other shrubs. Its toes had claws to help it dig in vegetation for tasty leaves and twigs. However, it was far from a helpless plant eater. It had very strong jaw muscles and could bite powerfully. In 1971, paleontologists found a now famous fossil of Protoceratops locked in a battle to the death with a Velociraptor (see p.83).

PERIOD	Late Cretaceous
FAMILY	Protoceratopsidae
DIET	herbivore
LENGTH	6 ft.
WEIGHT	500 lb.
FINDS	Asia

• March 27th •

TORVOSAURUS

A very large land predator, this "savage lizard" is known to be one of the most ferocious of the Jurassic predators. Its mouth was packed with fearsome teeth that were 4 inches long and shaped like blades. It had a heavy body, powerful hind legs, and short arms that finished in sharp claws like an eagle's talons.

PERIOD	Middle Jurassic
FAMILY	Megalosauridae
DIET	carnivore
LENGTH	33 ft.
WEIGHT	5 tons
FINDS	North America, Europe

• March 28th •

CERATONYKUS

Small and long-legged, this theropod appears to have adapted to running on the sands of a desert. It had tiny but strong front limbs with bird-like hands. It would have hunted small animals but also may have eaten insects.

PERIOD	Late Cretaceous
FAMILY	Alvarezsauridae
DIET	carnivore
LENGTH	2 ft.
WEIGHT	2 lb.
FINDS	Asia

• March 29th •

ELOPTERYX

First identified as a bird, this is in fact a bird-like dinosaur that has been much argued over by paleontologists. Its name means "marsh wing," and it lived and hunted for prey on marshy land in what is now Romania.

PERIOD	Late Cretaceous
FAMILY	Elopterygidae
DIET	carnivore
LENGTH	3 ft.
WEIGHT	44 lb.
FINDS	Europe

• March 30th •

FULGUROTHERIUM

This small plant eater, "lightning beast," was named after Lightning Ridge in New South Wales, Australia. It lived in a large herd and survived in an extreme environment. There would have been short, hot summers and very cold winters. It may even have spent the coldest months underground in a burrow.

PERIOD	Early Cretaceous
FAMILY	Hypsilophodontidae
DIET	herbivore
LENGTH	7 ft.
WEIGHT	24 lb.
FINDS	Australia

• March 31st •

BYRONOSAURUS

With its needle-like teeth, Byronosaurus would have tucked into prey such as lizards, frogs and snakes. It was a fast runner with a retractable talon on its second toe. Its large eyes and strong sense of smell helped it find its desert prey. Clusters of fossil nests partly buried under sand have been found.

PERIOD	Late Cretaceous
FAMILY	Troodontidae
DIET	carnivore
LENGTH	5 ft.
WEIGHT	9 lb.
FINDS	Asia

APRIL

• April 1st •

ALECTROSAURUS

This tyrannosaur had a large skull, tiny arms and strong legs. It lived in what is today's Gobi desert, but in the Late Cretaceous this area was a land of forests, lakes and streams. It hunted plant-eating dinosaurs such as Gilmoreosaurus (*see p.92*), killing them with its sharp teeth.

PERIOD	Late Cretaceous
FAMILY	Tyrannosauridae
DIET	carnivore
LENGTH	16 ft.
WEIGHT	1.1 tons
FINDS	Asia

• April 2ⁿᵈ •

PHILOVENATOR

Jaws full of sharp teeth and curved sickle claws on its feet were the perfect weapons for this fast-moving troodont. Its name means "love of the hunt" and it was a very successful predator. Its large eyes helped it hunt small mammals and lizards at dusk and in the dark shadows of the night.

PERIOD	Late Cretaceous
FAMILY	Troodontidae
DIET	carnivore
LENGTH	2 ft.
WEIGHT	3 lb.
FINDS	Asia

• April 3ʳᵈ •

NANKANGIA

The shape of this unusual oviraptor's jaw meant that it could not open its mouth very wide. This restricted what it could eat so, unlike other members of this family, it was a herbivore, probably eating only seeds and soft plant matter.

PERIOD	Late Cretaceous
FAMILY	Oviraptoridae
DIET	herbivore
LENGTH	7 ft.
WEIGHT	66 lb.
FINDS	Asia

• April 4th •
ANGATURAMA

Part of a family of "spined reptiles," Angaturama's crocodile-like jaws were armed with sharp teeth. This theropod would feast on pterosaurs and fish, as well as any other small animals it could catch or scavenge.

PERIOD	Early Cretaceous
FAMILY	Spinosauridae
DIET	carnivore
LENGTH	26 ft.
WEIGHT	1.1 tons
FINDS	South America

• April 5th •
MIRISCHIA

Quick-footed and agile, this biped snatched dragonflies from the air one minute and chased down small mammals on the ground the next. It lived in what is now Brazil.

PERIOD	Early Cretaceous
FAMILY	Compsognathidae
DIET	carnivore
LENGTH	7 ft.
WEIGHT	15 lb.
FINDS	South America

• April 6th •

PATAGOTITAN

One of the largest of them all, this titanosaur was a giant in a land of giants. As heavy as 12 African elephants, it stood 20 feet at the shoulder. Dinosaurs this big had lightweight hollow bones for movement. The bones were filled with air chambers that allowed oxygen to reach all parts of their enormous bodies.

PERIOD	Late Cretaceous
FAMILY	Titanosauridae
DIET	herbivore
LENGTH	130 ft.
WEIGHT	77.2 tons
FINDS	South America

• April 7th •

MEDUSACERATOPS

Many different species of armored plant eaters roamed in herds across the plains of what is now North America. This one had a particularly spectacular frill, and has been named after the Medusa of Greek myth, whose head was covered in snakes. It also had a sharp 3-foot-long horn over each eye.

PERIOD	Late Cretaceous
FAMILY	Ceratopsidae
DIET	herbivore
LENGTH	20 ft.
WEIGHT	2.2 tons
FINDS	North America

• April 8th •

CARNOTAURUS

A fast biped when chasing small but agile prey, Carnotaurus could reach speeds of up to 31 mph. It had a big head full of sharp teeth, tiny arms only 20 inches long and a great sense of smell to track prey. The two horns on the male's head could have been used to fight off another of its kind, maybe in a squabble over prey or a female.

PERIOD	Late Cretaceous
FAMILY	Abelisauridae
DIET	carnivore
LENGTH	30 ft.
WEIGHT	3.3 tons
FINDS	South America

• April 9th •

ORODROMEUS

With a name that means "mountain runner," this plant eater would have made a quick getaway if it spotted a Troodon (see p.16). It was small with a horny beak and grinding cheek teeth to eat fruits and tough plants. Its strong arms may have been used to dig burrows to live in.

PERIOD	Late Cretaceous
FAMILY	Parksosauridae
DIET	herbivore
LENGTH	8 ft.
WEIGHT	22 lb.
FINDS	North America

DINOSAUR WEAPONRY

Sharp teeth, long claws, spikes and horns – the weaponry of the dinosaur world was varied and vicious. Some weapons were used by predators to strike and kill, but many were used for defense against attack.

• April 10th •
EUOPLOCEPHALUS

Plant-eating ankylosaurs had a great way of defending themselves – a tail club. And Euoplocephalus needed both its bony tail and armor to protect itself against Gorgosaurus (see p.37).

PERIOD	Late Cretaceous
FAMILY	Ankylosauridae
DIET	herbivore
LENGTH	23 ft.
WEIGHT	2.2 tons
FINDS	North America

• April 11th •
STYRACOSAURUS

This was a peaceful animal traveling in large herds for protection against predators. If that did not work, it used its long horn and the spikes on its neck frill.

PERIOD	Late Cretaceous
FAMILY	Ceratopsidae
DIET	herbivore
LENGTH	18 ft.
WEIGHT	3 tons
FINDS	North America

• April 12th •
CARCHARODONTOSAURUS

A fearsome biped, this was one of the longest and heaviest of all the carnivores. Its strong jaws were full of serrated teeth up to 8 inches in length.

PERIOD	Early Cretaceous
FAMILY	Carcharodontosauridae
DIET	carnivore
LENGTH	46 ft.
WEIGHT	16.5 tons
FINDS	Africa

• April 13th •
TALOS

This bird-like troodont had sharp, curving talons on its feet that it held off the ground ready for action. It used them to capture prey, fight a rival or defend itself.

PERIOD	Late Cretaceous
FAMILY	Troodontidae
DIET	carnivore
LENGTH	7 ft.
WEIGHT	84 lb.
FINDS	North America

• April 14th •
ICHTHYOVENATOR

Using its sensitive nose to find fish in the water, this "fish hunter" grabbed prey with long jaws and sharp teeth. On land, it wielded its large thumb claws to slash prey such as small dinosaurs and pterosaurs.

PERIOD	Early Cretaceous
FAMILY	Spinosauridae
DIET	carnivore
LENGTH	10 ft.
WEIGHT	2.2 tons
FINDS	Asia

• April 15th •
KENTROSAURUS

This small stegosaur had an impressive array of armor. A double row of plates and spikes ran down its back to a tail that it swung in defense. The long shoulder spikes would have protected it from attacks from the side.

PERIOD	Late Jurassic
FAMILY	Stegosauridae
DIET	herbivore
LENGTH	15 ft.
WEIGHT	3.3 tons
FINDS	Africa

• April 16th •

BEIPIAOSAURUS

This extraordinary theropod was lightly built and a fast runner. The arrangement of long, stiff feathers on its head, back and tail is unusual. It may have been one of the earliest dinosaurs to have used feathers for display rather than warmth. It had large claws that would have been handy for grasping leaves to eat and useful weapons for defense.

PERIOD	Early Cretaceous
FAMILY	Therizinosauridae
DIET	herbivore
LENGTH	7 ft.
WEIGHT	185 lb.
FINDS	Asia

• April 17th •

YUTYRANNUS

A family group of these gigantic bipeds would have been a terrifying sight for any plant eater in its path. Yutyrannus – the name means "feathered tyrant" – was a distant cousin of Tyrannosaurus (see p.113). It was covered in fine, long feathers up to 8 inches in length for warmth, and had a large nose crest that was probably used for display to attract a mate.

PERIOD	Early Cretaceous
FAMILY	Tyrannosauridae
DIET	carnivore
LENGTH	30 ft.
WEIGHT	1.5 tons
FINDS	Asia

• April 18th •

BECKLESPINAX

The tall spines along this large predator's back created a curious hump or "sail" covered in feathers. This may have been used by the males to attract a female. Becklespinax had sharp claws that it used to kill prey such as Iguanodon (see p.27) or fight off rival theropod Baryonyx (see p.196).

PERIOD	Early Cretaceous
FAMILY	Allosauridae
DIET	carnivore
LENGTH	26 ft.
WEIGHT	1.7 tons
FINDS	Europe

• April 19th •

PARVICURSOR

One of the smallest dinosaurs yet found, this dinosaur's long legs were perfect for moving at a pace to escape the jaws of a predator. Its name means "slender runner." Parvicursor had the shortest of arms, with a sharp claw at the end of each one to dig for the termites and ants that it liked to eat.

PERIOD	Late Cretaceous
FAMILY	Alvarezsauridae
DIET	insectivore
LENGTH	2 ft.
WEIGHT	6 oz.
FINDS	Asia

• April 20th •

GRYPOSAURUS

Unusually, this very large hadrosaur did not have a crest. Its steeply rising beak opened into a mouth with an amazing 300 teeth inside for slicing up the plants it liked to eat. It normally walked on all fours but could stand on its hind legs to reach tasty leaves higher up.

PERIOD	Late Cretaceous
FAMILY	Hadrosauridae
DIET	herbivore
LENGTH	39 ft.
WEIGHT	5 tons
FINDS	North America

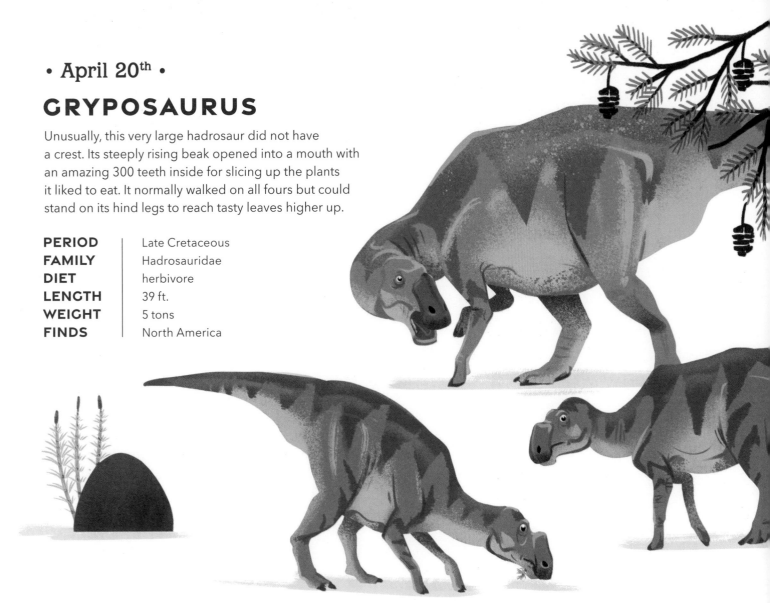

• April 21st •

SCIURUMIMUS

This "squirrel mimic" had a bushy tail that looked much like that of a tree squirrel today, although it was longer. A large-skulled carnivore, it hunted big prey, including other dinosaurs. Its young, on the other hand, caught insects and very small prey with slender, pointed teeth in the tips of their jaws.

PERIOD	Late Jurassic
FAMILY	Megalosauridae
DIET	carnivore
LENGTH	2 ft.
WEIGHT	9 lb.
FINDS	Europe

• April 22nd •

GUANLONG

A favorite snack for local allosaurs, this earliest known member of the tyrannosaur family was itself a hunter in the forest habitat where it lived. Guanlong had long arms and three-fingered clawed hands to grab and stab prey such as smaller dinosaurs, mammals and other small animals. Its nose crest was made from fused nasal bones and filled with air pockets. It was used to attract a mate.

PERIOD	Late Jurassic
FAMILY	Proceratosauridae
DIET	carnivore
LENGTH	10 ft.
WEIGHT	275 lb.
FINDS	Asia

• April 23rd •

TARCHIA

One of the largest ankylosaurs, Tarchia had a sturdy body supported by strong, stout legs. It was well defended by the sharp spikes all over its body and the flat club at the end of its tail, which it could swing at any large predator that threatened it. There was a network of air passages in its snout to moisten the dry air of the hot desert area in which it lived.

PERIOD	Late Cretaceous
FAMILY	Ankylosauridae
DIET	herbivore
LENGTH	28 ft.
WEIGHT	1.7 tons
FINDS	Asia

• April 24th •

CAUDIPTERYX

This turkey-sized dinosaur was small and quick-moving on long legs. Its short tail finished in a fan of feathers that were up to 6 inches long (its name means "tail feather"). Very lightweight, it had bones, including a wishbone, that were similar to those of a modern bird. It mainly ate plants, swallowing gizzard stones to grind them up in its stomach.

PERIOD	Early Cretaceous
FAMILY	Caudipteridae
DIET	omnivore
LENGTH	3 ft.
WEIGHT	15 lb.
FINDS	Asia

FISH-EATING DINOSAURS

Many dinosaurs lived along the coast or where a river joined the sea, and some of these were piscivores, or fish eaters, with a narrow snout and sharp teeth. A few were powerful swimmers, able to dive and pursue prey underwater.

• April 25th •
OXALAIA

This dinosaur had crocodile-like jaws. Its nostrils were set far back on its head so water would not get inside them when it was fishing.

PERIOD	Late Cretaceous
FAMILY	Spinosauridae
DIET	piscivore
LENGTH	46 ft.
WEIGHT	7.7 tons
FINDS	South America

• April 26th •
MASIAKASAURUS

Forward-pointing teeth at the front of its mouth helped this small biped spear fish in the water. This "vicious lizard" also ate snakes and mammals.

PERIOD	Late Cretaceous
FAMILY	Noasauridae
DIET	piscivore
LENGTH	7 ft.
WEIGHT	44 lb.
FINDS	Africa

• April 27th •
SIAMOSAURUS

Fishing mainly in the large lakes of an inland sea where today's Thailand is found, this dinosaur also tucked into small sauropods that it caught on land.

PERIOD	Early Cretaceous
FAMILY	Spinosauridae
DIET	piscivore
LENGTH	30 ft.
WEIGHT	3.3 tons
FINDS	Asia

• April 28th •
SUCHOMIMUS

Huge claws on each thumb helped this crocodile-like spinosaur pin down its prey. It hunted in the lagoons and river mouths of what is now Africa for fish, pterosaurs and small dinosaurs to eat.

PERIOD	Early Cretaceous
FAMILY	Spinosauridae
DIET	piscivore
LENGTH	36 ft.
WEIGHT	5.7 tons
FINDS	Africa

• April 29th •
OSTAFRIKASAURUS

Like other members of its family, Ostafrikasaurus, the oldest known spinosaur, lived a life half in and half out of lakes and rivers. It would have waded in to snap up fish and also seized prey on land that came to have a drink.

PERIOD	Late Jurassic
FAMILY	Spinosauridae
DIET	piscivore
LENGTH	33 ft.
WEIGHT	1.1 tons
FINDS	Africa

• April 30th •
HALSZKARAPTOR

About the size of a duck, this was an active diver and strong swimmer in pursuit of its prey. It had a long neck and razor-sharp claws in the flipper-like forelimbs that it used to propel itself through the water.

PERIOD	Late Cretaceous
FAMILY	Dromaeosauridae
DIET	piscivore
LENGTH	3 ft.
WEIGHT	3 lb.
FINDS	Asia

MAY

• May 1st •

BISTAHIEVERSOR

The enormous skull and deep snout housed large nostrils for sniffing out prey and jaws full of 64 sharp teeth to rip into it. This "destroyer" tyrannosaur was a top predator and stalked plant-eating dinosaurs such as Pentaceratops (see p.91) through forests and across rivers in what are now the badlands of the western United States. Like many of its relatives, this theropod had a relatively large brain for a dinosaur.

PERIOD	Late Cretaceous
FAMILY	Tyrannosauridae
DIET	carnivore
LENGTH	30 ft.
WEIGHT	3.3 tons
FINDS	North America

• May 2nd •
SPINOSAURUS

A crocodile-like fish eater, this dinosaur was at home in the water. It was adaptable, feeding on fish that it caught either standing in water or by swimming and diving after its prey. It probably fed on pterosaurs as well, and scavenged in the mangroves and along the shores and tidal flats where it lived.

PERIOD	Early Cretaceous
FAMILY	Spinosauridae
DIET	piscivore
LENGTH	52 ft.
WEIGHT	8.8 tons
FINDS	Africa

• May 3rd •
CITIPATI

This theropod had a long neck and short skull ending in its toothless beak. The tall crest on its head was for display to attract a mate. With large earth nests containing up to 30 eggs at a time, both parents kept the eggs warm and looked after the young.

PERIOD	Late Cretaceous
FAMILY	Oviraptoridae
DIET	omnivore
LENGTH	10 ft.
WEIGHT	185 lb.
FINDS	Asia

• May 4th •
PINACOSAURUS

A very well-protected dinosaur, it had a wide, stocky body with armored plates all the way down the back. The heavy club tail could deliver a deadly blow when swung at an attacker. It lived in a herd, traveling long distances to find plants to eat.

PERIOD	Late Cretaceous
FAMILY	Ankylosauridae
DIET	herbivore
LENGTH	16 ft.
WEIGHT	3.3 tons
FINDS	Asia

• May 5th •
ZHUCHENGTYRANNUS

Running on strong back legs after prey including duck-billed dinosaurs (see pp.50-51), this giant tyrannosaur was probably both a predator and a scavenger. Its powerful jaws were armed with 4-inch-long teeth that were serrated all the way down to their base.

PERIOD	Late Cretaceous
FAMILY	Tyrannosauridae
DIET	carnivore
LENGTH	36 ft.
WEIGHT	6.6 tons
FINDS	Asia

• May 6th •
PARALITITAN

This titanosaur had to eat plants all day long to feed its giant body, reaching high into the trees for the tastiest morsels. It lived alongside the Tethys ocean, in an area of tidal flats and mangrove swamps in what is now Egypt. It is one of the biggest sauropods yet known.

PERIOD	Middle Cretaceous
FAMILY	Titanosauridae
DIET	herbivore
LENGTH	89 ft.
WEIGHT	66.1 tons
FINDS	Africa

• May 7th •

ALBERTACERATOPS

The two large curved brow horns may have seen off threats from a large predator, but this small horned plant eater probably relied on camouflage for the best protection. Albertaceratops is thought to be one of the oldest members of a family that included Triceratops (see p.188).

PERIOD	Late Cretaceous
FAMILY	Ceratopsidae
DIET	herbivore
LENGTH	20 ft.
WEIGHT	3.9 tons
FINDS	North America

• May 8th •

VELOCIRAPTOR

Small and fast, with a vicious sickle-shaped claw on the second toe of each foot, this was a very efficient predator. Its flexible wrists and clawed hands helped it grab all kinds of prey smaller than itself as it chased it at speeds of up to 37 mph.

PERIOD	Late Cretaceous
FAMILY	Dromaeosauridae
DIET	carnivore
LENGTH	7 ft.
WEIGHT	44 lb.
FINDS	Asia

• May 9th •

SINOVENATOR

This bird-like troodont was a quick runner and had keen eyesight and hearing that helped it hunt down lizards and small mammals. It was the size of a chicken, but had many closely spaced serrated teeth as well as sharp sickle claws.

PERIOD	Early Cretaceous
FAMILY	Troodontidae
DIET	carnivore
LENGTH	3 ft.
WEIGHT	10 lb.
FINDS	Asia

FLYING REPTILES

The pterosaurs ruled the skies of the dinosaur world. However, they were not birds – they were flying reptiles. They soared on leathery wings, and some had long tails with flaps on the end to steer them through the air.

• May 10th •
GERMANODACTYLUS

About the size of a raven, this was one of the earliest fossil finds of a short-tailed pterosaur. It fed on small snails and other shelled creatures that it crushed with blunt teeth.

PERIOD	Late Jurassic
FAMILY	Germanodactylidae
DIET	piscivore
WINGSPAN	3 ft.
WEIGHT	4 lb.
FINDS	Europe

• May 11th •
TUPUXUARA

The splendid head crest on this short-tailed pterosaur would have given a vivid color display. It walked on all fours when it was on the ground.

PERIOD	Early Cretaceous
FAMILY	Thalassodromidae
DIET	omnivore
WINGSPAN	20 ft.
WEIGHT	51 lb.
FINDS	South America

• May 12th •
RHAMPHORHYNCHUS

The wing finger of this long-tailed pterosaur is probably the longest known to date. Its beak with its curved tip was used to snatch up fish, squid and insects.

PERIOD	Late Jurassic
FAMILY	Rhamphorhynchidae
DIET	piscivore
WINGSPAN	6 ft.
WEIGHT	2 lb.
FINDS	Europe

• May 13th •
PTERANODON

Flying in skies across the ancient world, this large, crested pterosaur had a small body but a large wingspan. Like other pterosaurs, it flapped its wings and glided through the air, using warm thermals to rise ever higher. With its toothless beak it skimmed up fish, squid and crabs without landing but walked on two legs when it was on the ground.

PERIOD	Late Cretaceous
FAMILY	Pteranodontidae
DIET	piscivore
WINGSPAN	33 ft.
WEIGHT	24 lb.
FINDS	North America, Europe, Asia

• May 14th •
DSUNGARIPTERUS

This large-headed pterosaur had curved jaws as well as a peculiar bony crest on its snout. When it walked on the shoreline, it would dip for mussels and worms.

PERIOD	Early Cretaceous
FAMILY	Dsungaripteridae
DIET	carnivore
WINGSPAN	11 ft.
WEIGHT	30 lb.
FINDS	Asia

• May 15th •
ANHANGUERA

When this eagle-eyed pterosaur took fish from the water, its curved teeth stopped the slippery prey from falling out of its mouth. Its name means "old devil."

PERIOD	Early Cretaceous
FAMILY	Anhangueridae
DIET	piscivore
WINGSPAN	15 ft.
WEIGHT	51 lb.
FINDS	South America, Africa

• May 16th •

QUETZALCOATLUS

This magnificent pterosaur may have been the largest flying animal ever to have existed. Unlike most pterosaurs, it lived inland, soaring on its long wings over lakes and pools, skim-feeding insects and fish from the surface. On land, it moved on all fours, hunting prey such as lizards and small dinosaurs.

PERIOD	Late Cretaceous
FAMILY	Azhdarchidae
DIET	carnivore
WINGSPAN	39 ft.
WEIGHT	550 lb.
FINDS	North America

• May 17th •

MAPUSAURUS

Living in family groups and hunting in packs, this was a very successful predator in what is now Argentina. It could have brought down and made short work of even the largest titanosaurs, such as Argentinosaurus (see p.99). Mapusaurus had a narrow skull and a jaw full of thin, blade-like teeth.

PERIOD	Late Cretaceous
FAMILY	Carcharodontosauridae
DIET	carnivore
LENGTH	43 ft.
WEIGHT	3.3 tons
FINDS	South America

• May 18th •

OVIRAPTOR

An agile and quick biped with a varied diet, this dinosaur had a hooked beak and could even break open bones. It was wrongly given a name that means "egg thief." Instead, it used its feathered body to keep its own eggs warm.

PERIOD	Late Cretaceous
FAMILY	Oviraptoridae
DIET	omnivore
LENGTH	5 ft.
WEIGHT	77 lb.
FINDS	Asia

• May 19th •

SCANSORIOPTERYX

The size of a pigeon, this feathered, bird-like dinosaur's name means "climbing wing." Its feet were specialized for perching and living in the trees. Its long third fingers helped it grasp the trunk as it moved up through the branches, and were used to dig out insects from under the bark to eat. It could not fly, but instead glided from branch to branch for short distances.

PERIOD	Middle Jurassic
FAMILY	Scansoriopterygidae
DIET	carnivore
LENGTH	1 ft.
WEIGHT	7 oz.
FINDS	Asia

• May 20th •

CHIROSTENOTES

This dinosaur could run up to 37 mph on its long, powerful legs after scuttling reptiles and mammals. It also had elongated second fingers to probe in the earth for grubs and frogs. Its parrot-like skull sported a bony crest and it had a well-developed sense of smell to find prey in the forest undergrowth.

PERIOD	Late Jurassic
FAMILY	Caenagnathidae
DIET	omnivore
LENGTH	5 ft.
WEIGHT	220 lb.
FINDS	North America

HORN-FACED DINOSAURS

Most members of the ceratopsid family sported armored skulls and dangerous horns. They were the rhinoceroses of the dinosaur world, and formidable when defending themselves.

• May 21st •
ANCHICERATOPS

The unusual neck frill was quite long and rectangular with bony knobs topped by backward-pointing spikes. It fed on swamp plants as it waded through water and mud.

PERIOD	Late Cretaceous
FAMILY	Ceratopsidae
DIET	herbivore
LENGTH	16 ft.
WEIGHT	2.2 tons
FINDS	North America

• May 22nd •
EINIOSAURUS

A downward-curving nose horn and bony ridges over the eyes are features of this ceratopsid. Living in a dry area, it needed specialized teeth for the tough plants.

PERIOD	Late Cretaceous
FAMILY	Ceratopsidae
DIET	herbivore
LENGTH	30 ft.
WEIGHT	5.5 tons
FINDS	North America

• May 23rd •
YEHUECAUHCERATOPS

This small "ancient horned face" had a short frill and two brow horns. It foraged for plants in a herd on the marshes and floodplains in what is now desert in Mexico.

PERIOD	Late Cretaceous
FAMILY	Ceratopsidae
DIET	herbivore
LENGTH	20 ft.
WEIGHT	4.4 tons
FINDS	North America

• May 24th •
PACHYRHINOSAURUS

At up to 9 feet in length, the frilled head shield topped one of the largest skulls of any known land animal. The thick knob of bone on the nose was probably for display.

PERIOD	Late Cretaceous
FAMILY	Ceratopsidae
DIET	herbivore
LENGTH	26 ft.
WEIGHT	4.4 tons
FINDS	North America

• May 25th •
PENTACERATOPS

With long horns above its eyes, a large one on its snout and two smaller ones in its cheeks, this "five-horned face" dinosaur lived in forests, snipping off leaves and fruit with its beak.

PERIOD	Late Cretaceous
FAMILY	Ceratopsidae
DIET	herbivore
LENGTH	26 ft.
WEIGHT	5.5 tons
FINDS	North America

• May 26th •
SINOCERATOPS

The magnificent frill on the head of this dinosaur had forward-curving hornlets on its rim. It also had a large nose horn on its long snout. This was the first ceratopsid found in today's China.

PERIOD	Late Cretaceous
FAMILY	Ceratopsidae
DIET	herbivore
LENGTH	23 ft.
WEIGHT	2.2 tons
FINDS	Asia

• May 27th •
FERGANOCEPHALE

The oldest known member of its family, this pachycephalosaur had the bulky body, short front legs and heavy tail of its relatives. The thick skull was used as a battering ram (see pp.150-151), either to fight for a mate, or to show strength in the herd.

PERIOD	Middle Jurassic
FAMILY	Pachycephalosauridae
DIET	herbivore
LENGTH	4 ft.
WEIGHT	200 lb.
FINDS	Asia

• May 28th •
GILMOREOSAURUS

An early duck-billed hadrosaur, fossils of this plant eater have been found in what is today's China. In prehistoric times, this area was covered with dense conifer forests, streams and lakes. A hadrosaur herd would have found plenty of plants to eat.

PERIOD	Late Cretaceous
FAMILY	Hadrosauridae
DIET	herbivore
LENGTH	26 ft.
WEIGHT	1.7 tons
FINDS	Asia

• May 29th •
CONCHORAPTOR

This small bird-like biped had a blunt snout and a large, strong beak. It could easily crush the shells of the crabs and mussels that it liked to eat. Its name means "plunderer of mussels."

PERIOD	Late Cretaceous
FAMILY	Oviraptoridae
DIET	carnivore
LENGTH	7 ft.
WEIGHT	22 lb.
FINDS	Asia

• May 30th •
HESPERONYCHUS

Using its excellent hearing and eyesight, this mini velociraptor (see p.83) hunted down the insects, lizards, small mammals and baby dinosaurs it liked to eat. It was one of the smallest predators, weighing about as much as a large chicken.

PERIOD	Late Cretaceous
FAMILY	Dromaeosauridae
DIET	carnivore
LENGTH	3 ft.
WEIGHT	4 lb.
FINDS	North America

• May 31st •
BRACHYCERATOPS

The "short-horned face" plant-eating dinosaur had a small nose horn and a medium-sized frill for protection. It would have been preyed on by the large meat eaters that shared its territory, such as Albertosaurus (see p.117).

PERIOD	Late Cretaceous
FAMILY	Ceratopsidae
DIET	herbivore
LENGTH	13 ft.
WEIGHT	175 lb.
FINDS	North America

JUNE

• June 1st •

SINOSAUROPTERYX

With its masked face, striped tail, dark back and pale underbelly, this tiny biped was well camouflaged in the dappled shade of the forest floor. Protected for the most part from larger predators, it hunted small mammals, lizards and insects to eat. This was the first dinosaur fossil ever found that showed evidence of having feathers.

PERIOD	Early Cretaceous
FAMILY	Compsognathidae
DIET	carnivore
LENGTH	4 ft.
WEIGHT	6 lb.
FINDS	Asia

• June 2nd •
TITANOCERATOPS
The earliest known member of its family, this dinosaur's name means "titanic horned face." It had a giant skull that was 9 feet long, a large frill, two curved brow horns, and a nose horn on its long snout. It weighed as much as an African elephant.

PERIOD	Late Cretaceous
FAMILY	Ceratopsidae
DIET	herbivore
LENGTH	22 ft.
WEIGHT	7.2 tons
FINDS	North America

• June 3rd •
SPHAEROTHOLUS
Its small size did not stop this plant eater from using its head as a battering ram. It swung its very round and thick skull at the soft parts of predators, or charged head-first at other males of its family in a fight for a female.

PERIOD	Late Cretaceous
FAMILY	Pachycephalosauridae
DIET	herbivore
LENGTH	7 ft.
WEIGHT	55 lb.
FINDS	North America

• June 4th •
ACHILLOBATOR

Like other raptors, this wily hunter had ferocious-looking sickle claws on its second toes. It was a fast-moving biped that hunted in a pack, leaping on prey such as the armored ankylosaur Talarurus. It attacked with the sickle claws and bit into its prey with sharp, serrated teeth. This was one of the largest of all the raptors.

PERIOD	Late Cretaceous
FAMILY	Dromaeosauridae
DIET	carnivore
LENGTH	20 ft.
WEIGHT	770 lb.
FINDS	Asia

• June 5th •
WULAGASAURUS

This duck-billed plant eater is one of many types of hadrosaur that lived in what is now northeastern China. It roamed in a family group, grazing on fruit and leaves from low-lying bushes and small trees.

PERIOD	Late Cretaceous
FAMILY	Hadrosauridae
DIET	herbivore
LENGTH	30 ft.
WEIGHT	3.3 tons
FINDS	Asia

LARGEST OF THEM ALL

In a world filled with enormous herbivores, the titanosaurs included some of the most gigantic beasts ever known. The name of the group comes from the Titans, the giants of ancient Greek myths and legends.

• June 6th •
TITANOSAURUS

The remains of a Titanosaurus, found earlier but named in 1877, were the first dinosaur bones to be discovered in India. Its long neck was supported by a massive trunk and topped by a small head.

PERIOD	Late Cretaceous
FAMILY	Titanosauridae
DIET	herbivore
LENGTH	120 ft.
WEIGHT	14.3 tons
FINDS	Asia

• June 7th •
MAMENCHISAURUS

It had a neck that made up half its body length, and its hind legs were shorter than its front legs. It may have used its whip-like tail to fend off predators.

PERIOD	Late Jurassic
FAMILY	Mamenchisauridae
DIET	herbivore
LENGTH	115 ft.
WEIGHT	33 tons
FINDS	China

• June 8th •
ANTARCTOSAURUS

The "southern lizard" used its peg-like teeth to snip off leaves. These dinosaurs did not chew, swallowing plants straight down in vast quantities.

PERIOD	Late Cretaceous
FAMILY	Titanosauridae
DIET	herbivore
LENGTH	98 ft.
WEIGHT	27.6 tons
FINDS	South America

• June 9th •

ARGENTINOSAURUS

"Argentina's lizard" held the record of largest dinosaur for many years. It would have taken around 40 years to reach adult size, with a height of 56 feet. This enabled it to reach the tallest conifers.

PERIOD	Late Cretaceous
FAMILY	Titanosauridae
DIET	herbivore
LENGTH	130 ft.
WEIGHT	110 tons
FINDS	South America

• June 10th •

SUPERSAURUS

This dinosaur was indeed a "super reptile." Even Allosaurus (see p.117) was not a threat to it. It ate leaves and shoots from the very top of forest trees.

PERIOD	Late Jurassic
FAMILY	Diplodocidae
DIET	herbivore
LENGTH	115 ft.
WEIGHT	38.6 tons
FINDS	North America

• June 11th •

FUTALOGNKOSAURUS

The "giant chief's" long neck was made up of 14 vertebrae and was more than 3 feet thick in places. One of the most complete giant sauropod skeletons was found in Argentina.

PERIOD	Late Cretaceous
FAMILY	Titanosauridae
DIET	herbivore
LENGTH	92 ft.
WEIGHT	77.2 tons
FINDS	South America

• June 12th •

ACHELOUSAURUS

Like so many in its family, this ceratopsid had a very thick armored skull, up to 5 feet long, with impressive horns. However, this was a peaceful herbivore that traveled in a herd, foraging for plants to eat as it moved across the plains of its home. The horns and frill fended off predators and were used for display. The frill may also have been flushed with blood in order to cool the dinosaur when needed.

PERIOD	Late Cretaceous
FAMILY	Ceratopsidae
DIET	herbivore
LENGTH	20 ft.
WEIGHT	3.3 tons
FINDS	North America

• June 13th •
SAURORNITHOIDES

This was a fast-running predator with excellent sight and hearing. Its large eyes allowed it to see in dim light, spotting with ease the small mammals and reptiles it liked to eat. It probably used its hands to seize prey and its sickle claws to kill it.

PERIOD	Late Cretaceous
FAMILY	Troodontidae
DIET	carnivore
LENGTH	10 ft.
WEIGHT	110 lb.
FINDS	Asia

• June 14th •
NQWEBASAURUS

The height of a chicken, this plant eater lived in what is now southern Africa. Fossils have been found with gastroliths in the stomach – stones it swallowed to help grind up tough plant food.

PERIOD	Early Cretaceous
FAMILY	Ornithomimidae
DIET	herbivore
LENGTH	3 ft.
WEIGHT	1 lb.
FINDS	Africa

• June 15th •
PEDOPENNA

A winged dinosaur that did not fly but may have been able to glide, this carnivore had a large claw on each foot to pin down prey.

PERIOD	Middle Jurassic
FAMILY	Anchiornithidae
DIET	carnivore
LENGTH	3 ft.
WEIGHT	2 lb.
FINDS	Asia

• June 16th •

MAIASAURA

In the late 1970s, in an area now known as "Egg Mountain" in Montana, USA, an amazing collection of Maiasaura fossil nests were found. It is known that this plant-eating dinosaur lived in very large herds, and the mud nests were only 23 feet apart and each contained 30 to 40 eggs about 6 inches across. The hatchlings would have stayed in the nests for up to a year.

PERIOD	Late Cretaceous
FAMILY	Hadrosauridae
DIET	herbivore
LENGTH	30 ft.
WEIGHT	4.4 tons
FINDS	North America

DINOSAURS WITH BEAKS

Many different dinosaurs had beaks made of keratin, the tough material found in the beaks of today's birds, as well as in nails, horns, claws and hooves. The beaks helped them bite and feed.

• June 17th •
HEYUANNIA

Remains of one of these beaked, toothless oviraptors were found with blue-green fossil eggs, arranged in overlapping circles. It was probably incubating them.

PERIOD	Late Cretaceous
FAMILY	Oviraptoridae
DIET	omnivore
LENGTH	5 ft.
WEIGHT	44 lb.
FINDS	Asia

• June 18th •
LIMUSAURUS

A young Limusaurus was born with teeth but lost them as it grew, forming a beak instead. It swallowed gastroliths, or stones, to help it digest tough plants.

PERIOD	Middle Jurassic
FAMILY	Noasauridae
DIET	herbivore
LENGTH	7 ft.
WEIGHT	33 lb.
FINDS	China

• June 19th •
FERRISAURUS

This parrot-beaked plant eater was around the size of a bighorn sheep and was a prey of large meat eaters such as Tyrannosaurus (see p.113).

PERIOD	Late Cretaceous
FAMILY	Leptoceratopsidae
DIET	herbivore
LENGTH	6 ft.
WEIGHT	330 lb.
FINDS	North America

• June 20th •
PSITTACOSAURUS

This small "parrot lizard" had strange horn-like growths on its cheeks. It used its horny, narrow beak to snip leaves off cycads, which it chewed with its cheek teeth.

PERIOD	Early Cretaceous
FAMILY	Psittacosauridae
DIET	herbivore
LENGTH	7 ft.
WEIGHT	44 lb.
FINDS	Asia

• June 21st •
AQUILOPS

Around the size of a cat, this horned dinosaur's skull was smaller than an adult human's hand. It had a strongly hooked beak with a strange extra bump on the front.

PERIOD	Early Cretaceous
FAMILY	Ceratopsidae
DIET	herbivore
LENGTH	2 ft.
WEIGHT	4 lb.
FINDS	North America

• June 22nd •
CHILESAURUS

Mysteriously, this dinosaur had a body and grasping hands that were very similar to the meat-eating predators of its time. However, it also had a small beak, and the jaws and teeth of a plant eater.

PERIOD	Late Jurassic
FAMILY	early sauropodmorph
DIET	herbivore
LENGTH	10 ft.
WEIGHT	220 lb.
FINDS	South America

• June 23rd •

AVACERATOPS

This small horn-faced plant eater was probably an earlier relative of Triceratops (see p.188). Like that dinosaur, Avaceratops had a short neck frill that was a solid sheet of bone with no fenestrae (openings covered with skin). It moved in a herd, feeding on ferns, cycads and low-growing conifers.

PERIOD	Late Cretaceous
FAMILY	Ceratopsidae
DIET	herbivore
LENGTH	14 ft.
WEIGHT	2.2 tons
FINDS	North America

• June 24th •

DROMICEIOMIMUS

Hollow-boned and lightweight, this emu-like dinosaur – its name means "emu mimic" – could run very fast, possibly up to 50 mph in short bursts! It had large eyes and good eyesight, able to spot the small insects, lizards and mammals it liked to eat in the bushes and undergrowth. It probably ate ginkgo and palm fruits as well as conifer cones when it found them.

PERIOD	Late Cretaceous
FAMILY	Ornithomimidae
DIET	omnivore
LENGTH	11 ft.
WEIGHT	220 lb.
FINDS	North America

• June 25th •

KILESKUS

This was not a big dinosaur, but it made up for that by being an efficient predator. It is one of the earliest known tyrannosaurs and less than half the size of Tyrannosaurus (see p.113). Kileskus roamed the forests of what is now western Siberia in Russia, hunting for mammals and other small prey. Predators, including large allosaurs and megalosaurs, would have been a danger it needed to avoid.

PERIOD	Middle Jurassic
FAMILY	Proceratosauridae
DIET	carnivore
LENGTH	17 ft.
WEIGHT	1,540 lb.
FINDS	Asia

• June 26th •

HAPLOCHEIRUS

On its long legs, this large early alvarezsaurid was able to move quickly to catch the small animals it liked to eat. It had sharp teeth and three-fingered clawed hands to grasp wriggling prey. Its bird-like features – feathers and clawed hands – predated other feathered dinosaurs by millions of years.

PERIOD	Late Jurassic
FAMILY	Alvarezsauridae
DIET	carnivore
LENGTH	7 ft.
WEIGHT	44 lb.
FINDS	Asia

• June 27th •

ANABISETIA

Feeding on all fours, this plant eater roamed in a herd through the tropical forests of what is today a desert landscape in South America. It was one of the smallest dinosaurs in that area, sharing its territory with some of the largest carnivores and herbivores that have ever lived.

PERIOD	Late Cretaceous
FAMILY	Hypsilophodontidae
DIET	herbivore
LENGTH	7 ft.
WEIGHT	44 lb.
FINDS	South America

• June 28th •

CHINDESAURUS

One of the fastest runners of the time on its long legs, this early theropod needed to move fast to catch prey and also to escape large reptiles such as Saurosuchus. It hunted small Triassic mammals, lizards and young dinosaurs. Its name comes from the Navajo word *chindi* meaning "ghost."

PERIOD	Late Triassic
FAMILY	Herrerasauridae
DIET	carnivore
LENGTH	8 ft.
WEIGHT	110 lb.
FINDS	North America

• June 29th •

FUKUIRAPTOR

This small allosaur is the most complete fossil theropod found in what is now Japan. As fearsome as other members of its family, it may have hunted in packs, stalking prey on long legs, and using its large claws and serrated teeth to injure and kill. Its name means "thief of Fukui."

PERIOD	Early Cretaceous
FAMILY	Megaraptoridae
DIET	carnivore
LENGTH	15 ft.
WEIGHT	385 lb.
FINDS	Asia

• June 30th •

GOBIVENATOR

Troodonts such as Gobivenator were small, quick-moving, agile predators with tails that made up most of their length. They also had clawed hands and fearsome sickle claws on their feet. Gobivenator – "Gobi desert hunter" – may have lived and hunted in a pack, chasing down small animals and sometimes baby dinosaurs.

PERIOD	Late Cretaceous
FAMILY	Troodontidae
DIET	carnivore
LENGTH	5 ft.
WEIGHT	2 lb.
FINDS	Asia

JULY

• July 1st •

AMPHICOELIAS

Wandering slowly across the plains in search of plants to eat, this was a large sauropod with a long neck and whip-like tail. It could reach high into the treetops for leaves or sweep its head across low-growing ferns for a tasty snack.

PERIOD	Late Jurassic
FAMILY	Diplodocidae
DIET	herbivore
LENGTH	82 ft.
WEIGHT	49.6 tons
FINDS	North America

• July 2nd •
CORYTHORAPTOR

This bird-like dinosaur looked like today's cassowary, with a casque, or helmet, on top of its head. Other members of the family had differently shaped casques. Corythoraptor lived in a hot and dry environment, eating drought-resistant plants, nuts and seeds, and topping up its diet with small lizards and mammals.

PERIOD	Late Cretaceous
FAMILY	Oviraptoridae
DIET	omnivore
LENGTH	7 ft.
WEIGHT	165 lb.
FINDS	Asia

• July 3rd •
BRACHYLOPHOSAURUS

Its name means "short-crested lizard," a description of this hadrosaur's unique bony crest that formed a flat shield on top of its skull. Compared with other duck-billed dinosaurs, it had quite a small head, a wide horny beak and unusually long front legs.

PERIOD	Late Cretaceous
FAMILY	Hadrosauridae
DIET	herbivore
LENGTH	36 ft.
WEIGHT	7.7 tons
FINDS	North America

• July 4th •

TYRANNOSAURUS

Perhaps the most famous dinosaur of all, this large predator had an enormous head with 4-foot-long jaws. Its excellent sense of smell helped it locate live or dead prey, from which it tore off chunks of meat with its 12-inch-long teeth, swallowing them whole because it could not chew. It would easily have killed a large Edmontosaurus (see pp.54–55).

PERIOD	Late Cretaceous
FAMILY	Tyrannosauridae
DIET	carnivore
LENGTH	41 ft.
WEIGHT	9.9 tons
FINDS	North America

• July 5th •
HUEHUECANAUHTLUS

Meaning "ancient duck," this hadrosaur's long name is taken from an Aztec dialect. Its fossils were found in what is now western Mexico, and revealed that the dinosaur had high spines along its backbone that resulted in a curved back. Like other members of its family, it was a herd animal.

PERIOD	Late Cretaceous
FAMILY	Hadrosauridae
DIET	herbivore
LENGTH	20 ft.
WEIGHT	1.7 tons
FINDS	North America

• July 6th •
MONOCLONIUS

The huge head was held close to the ground as it searched for plant food to grind down with its many cheek teeth. The short snout ended in a toothless beak and a single, sharp nose horn for protection against predators. It is likely that the frill, used for courtship, was larger in the males.

PERIOD	Late Cretaceous
FAMILY	Ceratopsidae
DIET	herbivore
LENGTH	16 ft.
WEIGHT	2.2 tons
FINDS	North America

• July 7th •

SAHALIYANIA

This hadrosaur lived at the very end of the Cretaceous in what is now northeast China. Walking on all fours, this crested duckbill was able to raise itself on two legs to reach tasty leaves. It could make deep, loud sounds by passing air through the crest to warn of danger, call to its herd or attract a mate.

PERIOD	Late Cretaceous
FAMILY	Hadrosauridae
DIET	herbivore
LENGTH	26 ft.
WEIGHT	2.4 tons
FINDS	Asia

• July 8th •

SAURORNITHOLESTES

An agile hunter with sickle claws on its feet, this small theropod belonged to a ferocious family. It probably hunted in a pack, killing other similar-sized dinosaurs as well as mammals and lizards. One is known to have caught a Quetzalcoatlus (see pp.86–87).

PERIOD	Late Cretaceous
FAMILY	Dromaeosauridae
DIET	carnivore
LENGTH	6 ft.
WEIGHT	22 lb.
FINDS	North America

MEAT-EATING DINOSAURS

To be a successful carnivore, a dinosaur needed to have long, powerful legs to run fast after their sometimes very nippy prey. To catch their food, they used long claws and strong jaws that housed a fearsome array of teeth.

• July 9th •
THANATOTHERISTES

This successful predator lived up to its name, which means "reaper of death." It is the oldest tyrannosaur from northern North America and was a predator of dinosaurs such as Xenoceratops (see p.31).

PERIOD	Late Cretaceous
FAMILY	Tyrannosauridae
DIET	carnivore
LENGTH	26 ft.
WEIGHT	2.2 tons
FINDS	North America

• July 10th •
EODROMAEUS

This "dawn runner" is one of the earliest carnivores found so far. It had sharp, curved teeth to eat prey, but was itself a favorite snack of the alligator-like reptiles of its time.

PERIOD	Late Triassic
FAMILY	early theropod
DIET	carnivore
LENGTH	4 ft.
WEIGHT	11 lb.
FINDS	South America

• July 11th •
HAGRYPHUS

A bird-like dinosaur, Hagryphus ate what it could find on the floodplain and in the peat swamps where it lived. Its diet included plants, small vertebrates and eggs.

PERIOD	Late Cretaceous
FAMILY	Caenagnathidae
DIET	omnivore
LENGTH	10 ft.
WEIGHT	130 lb.
FINDS	North America

• July 12th •
ALBERTOSAURUS

With more teeth than the larger tyrannosaurs – it had more than 60 – this predator would have crunched its way through the bone and flesh of its hadrosaur prey with ease.

PERIOD | Late Cretaceous
FAMILY | Tyrannosauridae
DIET | carnivore
LENGTH | 33 ft.
WEIGHT | 2.6 tons
FINDS | North America

• July 13th •
YANGCHUANOSAURUS

Possibly hunting in a pack, this powerful hunter had a massive tail about half its length. Its big but lightweight skull housed sharp, serrated teeth to kill sauropods such as Mamenchisaurus (see p. 98).

PERIOD | Middle Jurassic
FAMILY | Metriacanthosauridae
DIET | carnivore
LENGTH | 36 ft.
WEIGHT | 3.3 tons
FINDS | Asia

• July 14th •
ALLOSAURUS

This successful solitary hunter and scavenger was very large and may not have been able to run faster than 19 mph, so some fast, agile prey would have gotten away.

PERIOD | Late Jurassic
FAMILY | Allosauridae
DIET | carnivore
LENGTH | 39 ft.
WEIGHT | 2.8 tons
FINDS | Europe, North America

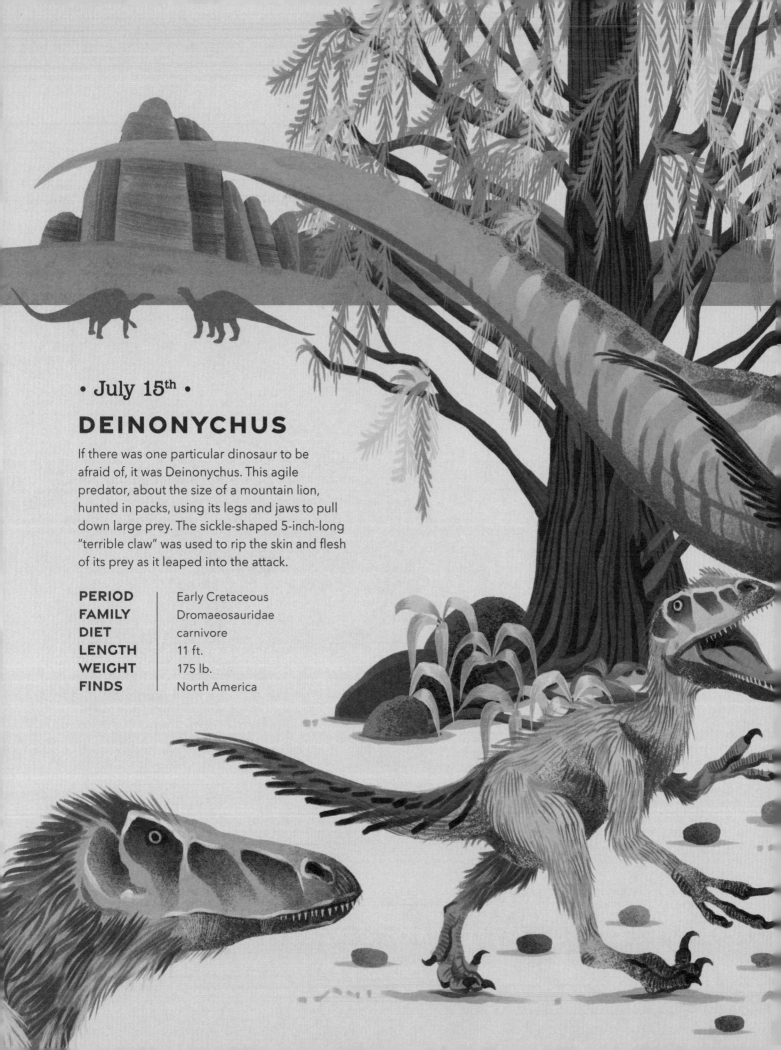

• July 15th •

DEINONYCHUS

If there was one particular dinosaur to be afraid of, it was Deinonychus. This agile predator, about the size of a mountain lion, hunted in packs, using its legs and jaws to pull down large prey. The sickle-shaped 5-inch-long "terrible claw" was used to rip the skin and flesh of its prey as it leaped into the attack.

PERIOD	Early Cretaceous
FAMILY	Dromaeosauridae
DIET	carnivore
LENGTH	11 ft.
WEIGHT	175 lb.
FINDS	North America

• July 16th •

TENONTOSAURUS

With a long stiff tail that was more than half its length, this plant eater had a flexible neck and horny beak to nip off the leaves it liked to eat. Moving in a herd gave this peaceful dinosaur some protection, but unfortunately it could not run fast enough to get away from a determined pack of Deinonychus.

PERIOD	Early Cretaceous
FAMILY	Hypsilophodontidae
DIET	herbivore
LENGTH	26 ft.
WEIGHT	2.2 tons
FINDS	North America

• July 17th •

RINCHENIA

Remains of this dinosaur were found in Mongolia. At first it was mistaken for an Oviraptor (*see p.88*) because it was similar in size and had a tall, domed crest. Later Rinchenia was identified as a different but close genus, or branch, of the family.

PERIOD	Late Cretaceous
FAMILY	Oviraptoridae
DIET	omnivore
LENGTH	8 ft.
WEIGHT	300 lb.
FINDS	Asia

• July 18th •

HUNGAROSAURUS

A medium-sized ankylosaur, this dinosaur had bony armor across its back and flanks to protect it from attack. It lived on floodplains, where it browsed on the low-growing plants.

PERIOD	Late Cretaceous
FAMILY	Nodosauridae
DIET	herbivore
LENGTH	13 ft.
WEIGHT	1,450 lb.
FINDS	Europe

• July 19th •

AUSTRORAPTOR

Although it was one of the largest in its family, Austroraptor had relatively small arms. Its narrow skull was full of conical, rather than serrated, teeth, with which it would have gripped prey such as small animals, fish and pterosaurs. This "southern plunderer" probably also scavenged on dead animals.

PERIOD	Late Cretaceous
FAMILY	Dromaeosauridae
DIET	carnivore
LENGTH	20 ft.
WEIGHT	770 lb.
FINDS	South America

• July 20th •

LANZHOUSAURUS

The lower jaw, or mandible, of this "Lanzhou lizard" was surprisingly large, at more than 3 feet in length. Its grinding teeth, at the back of its mouth, were also big, among the largest of any plant eater found so far. It is not known whether this was because it ate one particular type of plant.

PERIOD	Early Cretaceous
FAMILY	Iguanodontidae
DIET	herbivore
LENGTH	3 ft.
WEIGHT	6.6 tons
FINDS	Asia

SEA REPTILES

Living alongside the dinosaurs were some truly enormous reptiles in the seas and oceans. Plesiosaurus, Elasmosaurus and Pliosaurus were the largest in the Jurassic, while in the Cretaceous, mosasaurs dominated.

• July 21st •
KRONOSAURUS

With its streamlined body, long head and short neck, this was a fierce predator. It even ate smaller members of its own family!

PERIOD	Early Cretaceous
FAMILY	Pliosauridae
DIET	piscivore
LENGTH	36 ft.
WEIGHT	22 tons
FINDS	Australia, South America

• July 22nd •
PLIOSAURUS

Hunting mainly squid, ammonites and fish, with large teeth up to 1 foot in length, Pliosaurus could also tackle Plesiosaurus and ichthyosaurs.

PERIOD	Middle Jurassic
FAMILY	Pliosauridae
DIET	piscivore
LENGTH	43 ft.
WEIGHT	29.8 tons
FINDS	Europe

• July 23rd •
PLESIOSAURUS

Using its flippers to steer and brake, this reptile powered through the water after fish, squid and other sea creatures. It swallowed prey whole.

PERIOD	Early Jurassic
FAMILY	Plesiosauridae
DIET	piscivore
LENGTH	15 ft.
WEIGHT	1,100 lb.
FINDS	Europe

• July 24th •
MOSASAURUS

The deadliest of all sea creatures, this reptile is related to today's Komodo dragon (which lives on land). Mosasaurus ate sharks, fish, other mosasaurs and snatched pterosaurs from the air. An extra joint halfway along its jaw allowed it to swallow everything whole.

PERIOD	Late Cretaceous
FAMILY	Mosasauridae
DIET	piscivore
LENGTH	56 ft.
WEIGHT	33.1 tons
FINDS	Europe, North America

• July 25th •
TEMNODONTOSAURUS

This large deep-water ichthyosaur had eyes that were 8 inches across. When it dived, its big eyes were protected by thin bony plates called sclerotic rings.

PERIOD	Late Jurassic
FAMILY	Temnodontosauridae
DIET	piscivore
LENGTH	39 ft.
WEIGHT	7.7 tons
FINDS	Europe

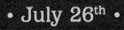

• July 26th •
ELASMOSAURUS

A small head topped what is possibly the longest neck of any animal – it had more than 70 vertebrae! It was probably an ambush predator of small fish.

PERIOD	Late Cretaceous
FAMILY	Elasmosauridae
DIET	piscivore
LENGTH	49 ft.
WEIGHT	9.9 tons
FINDS	North America, Asia

• July 27th •
TONGTIANLONG

This feathered dinosaur had a sharp toothless beak and a dome-shaped head topped by a crest of bone. The crest was used as a display to attract mates and scare away rivals.

PERIOD	Late Cretaceous
FAMILY	Oviraptoridae
DIET	omnivore
LENGTH	8 ft.
WEIGHT	660 lb.
FINDS	Asia

• July 28th •
QIAOWANLONG

Its heavy front legs and shoulders supported the 20-foot-long neck with which this titanosaur reached up for leaves. It lived in forests in what is now China.

PERIOD	Early Cretaceous
FAMILY	Euhelopodidae
DIET	herbivore
LENGTH	39 ft.
WEIGHT	9.9 tons
FINDS	Asia

• July 29th •
ANKYLOSAURUS

Built like a tank, with a club weighing 110 pounds at the end of its tail, this armored dinosaur could smash the teeth or head of most predators.

PERIOD	Late Cretaceous
FAMILY	Ankylosauridae
DIET	herbivore
LENGTH	33 ft.
WEIGHT	4.4 tons
FINDS	North America

• July 30th •

DICRAEOSAURUS

The long spines all down this plant eater's neck and back may have acted as a form of defense, making it difficult for a predator such as Megalosaurus (see p.57) to bite into it.

PERIOD	Late Jurassic
FAMILY	Dicraeosauridae
DIET	herbivore
LENGTH	46 ft.
WEIGHT	5.5 tons
FINDS	Africa

• July 31st •

BAHARIASAURUS

Although it was not the biggest predator in its habitat, this very large carnivore was a fierce hunter of smaller dinosaurs and other animals. It was a fast runner but may not have survived if chased by a Carcharodontosaurus (see p.68) in what is today's Egypt.

PERIOD	Late Cretaceous
FAMILY	Bahariasauridae
DIET	carnivore
LENGTH	39 ft.
WEIGHT	4.4 tons
FINDS	Africa

AUGUST

• August 1st •
ALETOPELTA

This ankylosaur's name means "wandering shield" and it was very well protected. Armored plates on its back, two large spikes above its shoulders, spikes along its sides and a heavy club on its tail would have seen off a lot of predators. It was a medium-sized member of the family, and had leaf-shaped teeth to break up the plants it ate.

PERIOD	Late Cretaceous
FAMILY	Ankylosauridae
DIET	herbivore
LENGTH	20 ft.
WEIGHT	2.2 tons
FINDS	North America

• August 2nd •
TURANOCERATOPS

Using its beak to pluck flowers and its hard jaw for tough vegetation, this "Turan horned face" munched on ferns, cycads and conifers. With two brow horns, but no nose horn, this is the first recorded ceratopsid from Asia and its remains were found in what is now Uzbekistan.

PERIOD	Late Cretaceous
FAMILY	Ceratopsidae
DIET	herbivore
LENGTH	7 ft.
WEIGHT	385 lb.
FINDS	Asia

• August 3rd •
ABRICTOSAURUS

This biped was an early member of its family, and it both hunted small animals and scavenged. It probably also dug for roots and snipped at plants with the cutting beak at the front of its mouth. It lived in an area that is now southern Africa.

PERIOD	Early Jurassic
FAMILY	Heterodontosauridae
DIET	omnivore
LENGTH	4 ft.
WEIGHT	100 lb.
FINDS	Africa

• August 4th •
XIAOSAURUS

Lightly built and fast on its two four-toed feet, the name Xiaosaurus means the "dawn lizard." It lived in a large family group that moved through the forests to find low-lying plants to eat while avoiding predators.

PERIOD	Middle Jurassic
FAMILY	Fabrosauridae
DIET	herbivore
LENGTH	4 ft.
WEIGHT	15 lb.
FINDS	Asia

• August 5th •
NOTHRONYCHUS

A strange sight, Nothronychus had a small head, a long, thin neck, and arms that ended in sharply clawed hands. Its beak and leaf-shaped teeth were used to shred plant matter in the tropical jungle in which it lived. It is related to the tyrannosaurs, but this branch of the family were plant eaters.

PERIOD	Late Cretaceous
FAMILY	Therizinosauridae
DIET	herbivore
LENGTH	20 ft.
WEIGHT	1.3 tons
FINDS	North America

• August 6th •
ACRISTAVUS

Unlike most other hadrosaurs (see pp.50-51), this dinosaur did not have an elaborate headdress – its name means "crestless grandfather." It lived near rivers and lakes on the floodplains of the Western Interior Seaway in what is now North America.

PERIOD	Late Cretaceous
FAMILY	Hadrosauridae
DIET	herbivore
LENGTH	28 ft.
WEIGHT	2.2 tons
FINDS	North America

• August 7th •

HOMALOCEPHALE

Like other pachycephalosaurs, this dinosaur had a thickened head for head-butting contests or to hit an attacker in defense (see pp.150-151). Its large eyes were good for spotting danger and long legs would help it to run away fast. It fed mainly on leaves and seeds that its family group searched for in the high-altitude forests that covered what is Mongolian desert today.

PERIOD	Late Cretaceous
FAMILY	Homalocephalidae
DIET	herbivore
LENGTH	10 ft.
WEIGHT	440 lb.
FINDS	Asia

• August 8th •

SUCHOSAURUS

The crocodile-like snout and many sharp teeth were very efficient for catching fish prey in the river deltas where Suchosaurus hunted. It also scavenged for and hunted small animals on land. Its name means "crocodile lizard" because it was mistaken for a crocodile when its fossils were first found.

PERIOD	Early Cretaceous
FAMILY	Spinosauridae
DIET	piscivore
LENGTH	30 ft.
WEIGHT	2.2 tons
FINDS	Europe

• August 9th •

TETHYSHADROS

This duck-billed dinosaur is one of the most complete fossils discovered so far. It lived on a small island in the middle of the Tethys ocean, the prehistoric sea that separated what is today's Africa and Europe. The hadrosaur is smaller than most of its family, probably because there was limited plant food for it to eat on the island.

PERIOD	Late Cretaceous
FAMILY	Hadrosauridae
DIET	herbivore
LENGTH	13 ft.
WEIGHT	770 lb.
FINDS	Europe

FEATHERED DINOSAURS

Recent discoveries have shown that many non-bird dinosaurs were covered or partially covered in feather-like structures. Most examples of this have been seen in theropods, the meat-eating dinosaurs.

• August 10th •
SINORNITHOSAURUS

One of the first dinosaurs discovered with feathers, this small "Chinese bird-lizard" did not fly. Instead it leaped with precision from branch to branch in pursuit of small animals.

PERIOD	Early Cretaceous
FAMILY	Dromaeosauridae
DIET	carnivore
LENGTH	4 ft.
WEIGHT	20 lb.
FINDS	Asia

• August 11th •
KULINDADROMEUS

This long-legged early feathered plant eater was about the size of a turkey. It lived in a land of lakes and volcanoes in what is today's Siberia.

PERIOD	Middle Jurassic
FAMILY	early ornithopod
DIET	herbivore
LENGTH	5 ft.
WEIGHT	31 lb.
FINDS	Asia

• August 12th •
CAIHONG

This crow-sized dinosaur had iridescent feathers – its name means "rainbow." It glided through the forests on four wings after small mammals, lizards and insects.

PERIOD	Late Jurassic
FAMILY	Anchiornithidae
DIET	carnivore
LENGTH	16 in.
WEIGHT	1 lb.
FINDS	Asia

• August 13th •
ANCHIORNIS

With its startling feathered crest, this chicken-sized dinosaur would have been easy to spot. It did not fly despite its wings being like those of a modern bird.

PERIOD	Late Jurassic
FAMILY	Anchiornithidae
DIET	carnivore
LENGTH	16 in.
WEIGHT	7 oz.
FINDS	Asia

• August 14th •
TIANYURAPTOR

Although it was swift on its feet and could speed after small animal prey, this medium-sized dromaeosaur was not able to glide like later members of its family. Its name means "Tianyu thief."

PERIOD	Early Cretaceous
FAMILY	Dromaeosauridae
DIET	carnivore
LENGTH	5 ft.
WEIGHT	9 lb.
FINDS	Asia

• August 15th •
XIAOTINGIA

This pigeon-sized dinosaur, with its feathers, wishbone and long arms, may be one of the earliest examples of a bird, although scientists are still arguing about this.

PERIOD	Late Jurassic
FAMILY	Anchiornithidae
DIET	insectivore
LENGTH	20 in.
WEIGHT	3 lb.
FINDS	Asia

• August 16th •

ARCHAEOPTERYX

In Jurassic times, this part of what is now Germany was a series of islands surrounded by shallow seas and lagoons. This is where Archaeopteryx, possibly the earliest known ancestor of today's birds, lived. It was able to glide, and used three claws on each wing to climb trees to escape danger or to hunt small prey.

PERIOD	Late Jurassic
FAMILY	Archaeopterygidae
DIET	omnivore
LENGTH	20 in.
WEIGHT	4 lb.
FINDS	Europe

• August 17th •

WELLNHOFERIA

Closely related to Archaeopteryx, this dinosaur was a bit larger and had a shorter tail, more like a modern bird. It lived in and around the forests and lagoons of western Europe, where it hunted for the insects and small animals it liked to eat.

PERIOD	Late Jurassic
FAMILY	Archaeopterygidae
DIET	omnivore
LENGTH	2 ft.
WEIGHT	3 lb.
FINDS	Europe

• August 18th •
CHARONOSAURUS

The enormous hollow crest curving backward from the top of the skull was used to make loud, rumbling calls. The dinosaur breathed in air that passed through the crest like air passes through a musical wind instrument. The calls would have been to warn of danger, to attract and defend mates, or simply to communicate with other dinosaurs in the herd.

PERIOD	Late Cretaceous
FAMILY	Hadrosauridae
DIET	herbivore
LENGTH	43 ft.
WEIGHT	7.2 tons
FINDS	Asia

• August 19th •
FABROSAURUS

Living in a family group for protection, this tiny, lightly built, early plant eater had a horn-covered beak and small, sharp teeth. It could move quickly on two legs to escape predators.

PERIOD	Early Jurassic
FAMILY	Fabrosauridae
DIET	herbivore
LENGTH	3 ft.
WEIGHT	84 lb.
FINDS	Africa

• August 20th •
GASOSAURUS

Fossils of this hunter were found by a Chinese gas-mining company, so it was given a name that means "gas lizard." It had ferocious claws, large jaws with sharp teeth and a stiff, pointed tail, which it would have whipped round as a weapon if needed.

PERIOD	Middle Jurassic
FAMILY	early theropod
DIET	carnivore
LENGTH	13 ft.
WEIGHT	550 lb.
FINDS	Asia

• August 21st •
LYTHRONAX

The oldest tyrannosaur found so far, this theropod had a wide skull to allow for very powerful jaw muscles. Its eyes pointed fully forward like an eagle's or snake's do today. This helped it pinpoint prey such as a hadrosaur or ankylosaur that it then crunched with its large, curved teeth.

PERIOD	Late Cretaceous
FAMILY	Tyrannosauridae
DIET	carnivore
LENGTH	26 ft.
WEIGHT	2.8 tons
FINDS	North America

• August 22nd •

AEPYORNITHOMIMUS

A fast-moving feathered dinosaur with slender, long toes, this theropod lived in a windswept desert. It could not fly but used its "wings" for communication or display.

PERIOD	Late Cretaceous
FAMILY	Ornithomimidae
DIET	herbivore
LENGTH	3 ft.
WEIGHT	66 lb.
FINDS	Asia

• August 23rd •

CHASMOSAURUS

The elongated frill of this ceratopsid could be flushed with blood to allow the dinosaur to cool down. It was a medium-sized plant eater that traveled in a large herd for protection.

PERIOD	Late Cretaceous
FAMILY	Ceratopsidae
DIET	herbivore
LENGTH	18 ft.
WEIGHT	2.8 tons
FINDS	North America

• August 24th •

TOROSAURUS

This plant eater would have used its impressively long brow horns to fight back if a tyrannosaur threatened its young or family. It had one of the largest skulls of any land animal ever.

PERIOD	Late Cretaceous
FAMILY	Ceratopsidae
DIET	herbivore
LENGTH	30 ft.
WEIGHT	6.6 tons
FINDS	North America

• August 25th •

DELAPPARENTIA

Wandering in a herd through the woodlands, mudflats and coastal marshes of what is today's Spain, this iguanodont browsed on low plants and bushes. It used its large thumb spikes to defend itself against attacks from predators such as Baryonyx (see p.196).

PERIOD	Early Cretaceous
FAMILY	Iguanodontidae
DIET	herbivore
LENGTH	33 ft.
WEIGHT	3.9 tons
FINDS	Europe

SMALL HERDERS

If you were a small dinosaur, you had every reason to travel in a large group. Being surrounded by many members of your family, you could eat, nest and travel long distances in relative safety from predators.

• August 26th •
HYPSILOPHODON

Living on the wooded plains and shores of what is now southern England, this fast-moving biped traveled in large family groups or with a big herd.

PERIOD	Early Cretaceous
FAMILY	early ornithopod
DIET	herbivore
LENGTH	5 ft.
WEIGHT	53 lb.
FINDS	Europe

• August 27th •
OTHNIELOSAURUS

Small family herds of this "dwarf lizard" were a common sight during the Late Jurassic across what is now central North America.

PERIOD	Late Jurassic
FAMILY	Nanosauridae
DIET	herbivore
LENGTH	7 ft.
WEIGHT	22 lb.
FINDS	North America

• August 28th •
QANTASSAURUS

This was a fast biped with long legs. Named after the Australian airline Qantas, this plant eater traveled in small herds and possibly spent cold winter months in a burrow.

PERIOD	Early Cretaceous
FAMILY	early iguanodont
DIET	herbivore
LENGTH	7 ft.
WEIGHT	100 lb.
FINDS	Australia

• August 29th •
GASPARINISAURA

One of the first of its family to be unearthed in today's Argentina, this very small herder was found with gastroliths (stomach stones) to grind up the tough plant material it ate.

PERIOD	Late Cretaceous
FAMILY	early iguanodont
DIET	herbivore
LENGTH	5 ft.
WEIGHT	77 lb.
FINDS	South America

• August 30th •
DRYOSAURUS

Preyed on by large theropods such as Allosaurus (*see p.117*), a herd of Dryosaurus needed to move fast in the forests and on the plains where it lived.

PERIOD	Late Jurassic
FAMILY	Dryosauridae
DIET	herbivore
LENGTH	14 ft.
WEIGHT	200 lb.
FINDS	Africa, North America

• August 31st •
PARKSOSAURUS

For extra protection of the herd, this dinosaur could hear low-frequency sounds and had excellent eyesight, both of which helped it spot the approach of a predator.

PERIOD	Late Cretaceous
FAMILY	Parksosauridae
DIET	herbivore
LENGTH	8 ft.
WEIGHT	110 lb.
FINDS	North America

SEPTEMBER

• September 1st •
ACROTHOLUS

A high, dome-shaped casque, or helmet, of solid bone more than 4 inches thick crowned the skull of this dog-sized biped. It is one of the earliest known members of its bone-headed family, foraging in a herd for the mosses and ferns it liked to eat in the woodlands of what is today's Canada.

PERIOD	Late Cretaceous
FAMILY	Pachycephalosauridae
DIET	herbivore
LENGTH	6 ft.
WEIGHT	88 lb.
FINDS	North America

• September 2nd •

AFROVENATOR

With long, powerful legs and a stiff tail as counterbalance, this "African hunter" moved quickly after sauropod prey. It had powerful arms ending in clawed fingers and a mouthful of blade-like teeth up to 2 inches long.

PERIOD	Middle Jurassic
FAMILY	Megalosauridae
DIET	carnivore
LENGTH	26 ft.
WEIGHT	1.3 tons
FINDS	Africa

• September 3rd •

IRRITATOR

The long, crocodile-like skull of this spinosaur was perfect for fishing in the river deltas where it lived. It also scavenged on land and caught pterosaurs.

PERIOD	Early Cretaceous
FAMILY	Spinosauridae
DIET	carnivore
LENGTH	26 ft.
WEIGHT	2.2 tons
FINDS	South America

• September 4th •

MAGNOSAURUS

A medium-sized predator, Magnosaurus ran fast on two legs after small prey. Like other members of its family, it would also have scavenged on any dead animals it found.

PERIOD	Middle Jurassic
FAMILY	Megalosauridae
DIET	carnivore
LENGTH	13 ft.
WEIGHT	1,100 lb.
FINDS	Europe

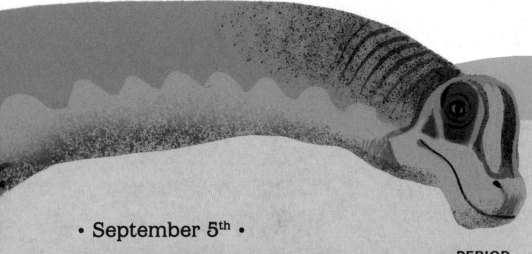

• September 5th •

DREADNOUGHTUS

Weighing as much as six African elephants, this huge titanosaur had an 36-foot-long neck balanced by a 30-foot-long tail. It had to eat all the time to get enough fuel for its body in the temperate forests of what is now the tip of South America.

PERIOD	Late Cretaceous
FAMILY	Diplodocidae
DIET	herbivore
LENGTH	85 ft.
WEIGHT	39.7 tons
FINDS	South America

ROOF LIZARDS

The plant-eating stegosaurs had large bony plates along their backs that may have warned off predators or allowed the members of their families to recognize them. The name Stegosaurus is Greek for "roof lizard."

• September 6th •
ADRATIKLIT

It was the earliest armored dinosaur find in Morocco, Africa, and may be the oldest stegosaur found worldwide. This plant eater's name comes from the local Berber words that mean "mountain lizard."

PERIOD	Middle Jurassic
FAMILY	Stegosauridae
DIET	herbivore
LENGTH	30 ft.
WEIGHT	4.4 tons
FINDS	Africa

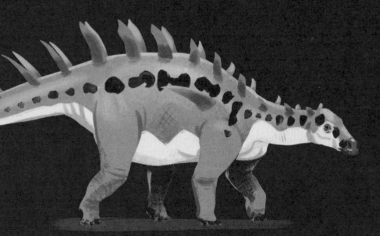

• September 7th •
PARANTHODON

This was the first African dinosaur to be discovered. Its remains were found in South Africa in 1845. It had spikes running from its neck to the tip of its tail.

PERIOD	Early Cretaceous
FAMILY	Stegosauridae
DIET	herbivore
LENGTH	16 ft.
WEIGHT	2.2 tons
FINDS	Africa

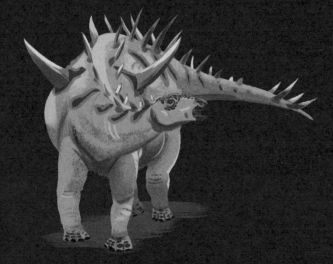

• September 8th •
HUAYANGOSAURUS

This early stegosaur was unusual in that it had 14 teeth at the front of its snout, unlike its later relatives. Its spiky armor and shoulder spines would have certainly protected it from predators.

PERIOD	Middle Jurassic
FAMILY	Huayangosauridae
DIET	herbivore
LENGTH	15 ft.
WEIGHT	1,850 lb.
FINDS	Asia

• September 9th •

HESPEROSAURUS

This armored dinosaur had another method of defense. It would have swung its tail very effectively as a weapon against fierce predators such as Ceratosaurus (see p.19).

PERIOD	Late Jurassic
FAMILY	Stegosauridae
DIET	herbivore
LENGTH	20 ft.
WEIGHT	2.8 tons
FINDS	North America

• September 10th •

GIGANTSPINOSAURUS

The huge shoulder spikes meant that this relatively small stegosaur could not be ignored. The spikes were probably used for display as well as defense.

PERIOD	Late Jurassic
FAMILY	Stegosauridae
DIET	herbivore
LENGTH	15 ft.
WEIGHT	495 lb.
FINDS	Asia

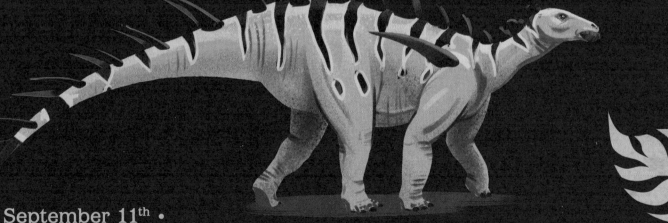

• September 11th •

DACENTRURUS

A heavily built dinosaur, Dacentrurus had spikes all over the place! And the tail spike had sharp cutting edges on the front and rear, which would have inflicted the maximum damage on any attacker.

PERIOD	Late Jurassic
FAMILY	Stegosauridae
DIET	herbivore
LENGTH	26 ft.
WEIGHT	6.1 tons.
FINDS	Europe

• September 12th •
FOSTEROVENATOR

This ceratosaur shared its habitat with some truly threatening larger theropods, such as Torvosaurus (see p.60) and Allosaurus (see p.117), and would have used speed to escape danger. However, it probably benefitted from them as well by scavenging meat from their kills.

PERIOD	Late Jurassic
FAMILY	Ceratosauridae
DIET	carnivore
LENGTH	8 ft.
WEIGHT	440 lb.
FINDS	North America

• September 13th •
OLOROTITAN

The unusual fan-shaped crest of this hadrosaur was brightly colored to attract a mate. It was hollow to allow the plant eater to make the hooting sounds it used to communicate with its herd. The name means "swan giant" because the dinosaur had 18 vertebrae in its neck, three more than any other hadrosaur.

PERIOD	Late Cretaceous
FAMILY	Hadrosauridae
DIET	herbivore
LENGTH	39 ft.
WEIGHT	5 tons
FINDS	Asia

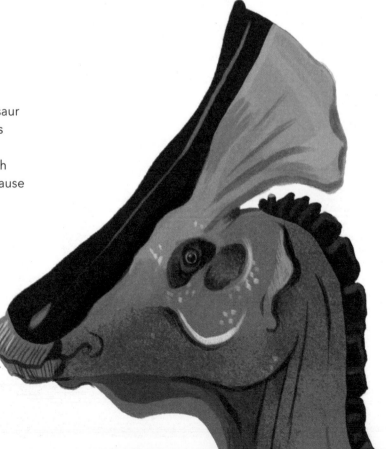

• September 14th •

MANIDENS

This pigeon-sized omnivore had long tusk-like teeth in both its upper and lower beaks. Its toes were unusually long and ended in curved, narrow claws much like those of modern birds. It had the ability to climb trees and is the earliest example so far of any dinosaur that could do this.

PERIOD	Middle Jurassic
FAMILY	Heterodontosauridae
DIET	omnivore
LENGTH	2 ft.
WEIGHT	1 lb.
FINDS	South America

• September 15th •

JEHOLOSAURUS

Relying on keen eyesight and good hearing to pick up signs of danger, Jeholosaurus was quick on its feet and agile in avoiding predators. It mainly ate plants that it ground up with its cheek teeth. However, it had strong front teeth, so may have also taken small animals or scavenged.

PERIOD	Early Cretaceous
FAMILY	Parksosauridae
DIET	herbivore
LENGTH	7 ft.
WEIGHT	330 lb.
FINDS	Asia

• September 16th •
PACHYCEPHALOSAURUS

Two "thick-headed lizards" clash in a head-butting competition for a female. This behavior is similar to the way today's musk oxen act during the mating season. These pachycephalosaurs were gentle at other times of the year, browsing on flowering shrubs and ferns in subtropical forested swamps.

PERIOD	Late Cretaceous
FAMILY	Pachycephalosauridae
DIET	herbivore
LENGTH	15 ft.
WEIGHT	990 lb.
FINDS	North America

• September 17th •

APATORAPTOR

With long legs for wading and running, this dinosaur roamed tropical swamplands looking for the crustaceans, fish and amphibians it liked to eat. It could not fly but probably used the feathers on its short arms for display.

PERIOD	Late Cretaceous
FAMILY	Caenagnathidae
DIET	omnivore
LENGTH	7 ft.
WEIGHT	395 lb.
FINDS	North America

• September 18th •

SPINOPS

"Spine face" had a large nose horn and two brow horns, as well as two long spikes at the top of its frill, and unusually, two forward-curving hooks in the middle. This cousin of Triceratops (see p.188) was a medium-sized member of the family. It lived in woodlands in what is now Canada, foraging for plants to eat with other members of its herd.

PERIOD	Late Cretaceous
FAMILY	Ceratopsidae
DIET	herbivore
LENGTH	23 ft.
WEIGHT	2.2 tons
FINDS	North America

• September 19th •

TRINISAURA

Fossils of this small, agile plant eater were found on what is now James Ross Island, Antarctica. When it lived there, the area was ice-free, with varied seasons and a rainy climate.

PERIOD	Late Cretaceous
FAMILY	early ornithopod
DIET	herbivore
LENGTH	10 ft.
WEIGHT	44 lb.
FINDS	Antarctica

• September 20th •

GOBISAURUS

Unlike later members of its family, this early ankylosaur did not have a club at the end of its tail. Its body armor would have protected it against large predators such as Shaochilong.

PERIOD	Early Cretaceous
FAMILY	Ankylosauridae
DIET	herbivore
LENGTH	20 ft.
WEIGHT	1,750 lb.
FINDS	Asia

EARLY BIRDS

The earliest birds arose from theropod dinosaurs in the Jurassic (*see pp. 134–135*). However, during the Cretaceous, a wide range of birds appeared, many becoming increasingly efficient at flying.

• September 21st •
IBEROMESORNIS

The size of a sparrow, this bird had a short tail bone but long tail feathers. Its feet were well-adapted for perching on the branches of its treetop habitat in what is now Spain.

PERIOD	Early Cretaceous
FAMILY	Iberomesornithidae
DIET	insectivore
WINGSPAN	8 in.
WEIGHT	0.7 oz
FINDS	Europe

• September 22nd •
HESPERORNIS

A large, fish-eating diver with tiny wings and legs near its tail, Hesperornis lumbered along on land. But in the water, it was a fast and agile predator.

PERIOD	Late Cretaceous
FAMILY	Hesperornithidae
DIET	piscivore
WINGSPAN	6 ft.
WEIGHT	20 lb.
FINDS	North America, Europe

• September 23rd •
ICHTHYORNIS

Its name means "fish bird" and it looked like some seabirds alive today. However, like its ancestors, its beak was full of sharp teeth. It could have remained in the air for long periods over water.

PERIOD	Late Cretaceous
FAMILY	Ichthyornithidae
DIET	piscivore
WINGSPAN	2 ft.
WEIGHT	14 oz.
FINDS	North America

• September 24th •
CRATOAVIS

This colorful songbird flitted from tree to tree in the tropical woodlands of what is now Brazil. Its recent discovery was the first time that a complete skeleton of an Early Cretaceous bird was found in South America.

PERIOD	Early Cretaceous
FAMILY	Enantiornithes
DIET	insectivore
WINGSPAN	4 in.
WEIGHT	1.4 oz
FINDS	South America

• September 25th •
YANORNIS

A strong flier with large wings, Yanornis probably fed on whatever small fish it could catch as it foraged in the shallows. Gastroliths in its stomach ground up the food.

PERIOD	Early Cretaceous
FAMILY	Songlingornithidae
DIET	piscivore
WINGSPAN	3 ft.
WEIGHT	2 lb.
FINDS	Asia

• September 26th •
GANSUS

This pigeon-sized bird used its webbed feet to push itself along in rivers and lakes, plunging its head or body below the surface to find fish and snails to eat.

PERIOD	Late Cretaceous
FAMILY	early Ornithurae
DIET	piscivore
WINGSPAN	16 in.
WEIGHT	4 oz.
FINDS	Asia

• September 27th •

LUANCHUANRAPTOR

A small, bird-like dinosaur, this "Luanchuan thief" relied on its speed and the sickle claws on its feet to defend itself against predators on the floodplains where it lived. It preyed on small animals and may have fished. Like other members of its family, such as Velociraptor (see p.83), it probably also hunted in a pack to overwhelm larger dinosaurs.

PERIOD	Late Cretaceous
FAMILY	Dromaeosauridae
DIET	carnivore
LENGTH	6 ft.
WEIGHT	10 lb.
FINDS	Asia

• September 28th •

SHUVUUIA

About the size of a turkey, Shuvuuia had short arms that each ended in a single large hooked claw and two other short digits. The claw could see off predators, but was more often used to tear open tree bark or ant hills to find insects to eat with its pointed beak full of tiny teeth. It was a fast runner, which helped it escape the larger dinosaurs that shared its home.

PERIOD	Late Cretaceous
FAMILY	Alvarezsauridae
DIET	insectivore
LENGTH	3 ft.
WEIGHT	6 lb.
FINDS	Asia

• September 29th •

LAQUINTASAURA

Around the size of a red fox, this early dinosaur was a plant eater, lightweight and a fast runner. Fossils of several of these dinosaurs have been found in a bone bed in what is today's western Venezuela, providing evidence that Laquintasaura lived in a herd.

PERIOD	Early Jurassic
FAMILY	Ornithischia
DIET	herbivore
LENGTH	3 ft.
WEIGHT	11 lb.
FINDS	South America

• September 30th •

APPALACHIOSAURUS

Living in densely wooded rainforests, this tyrannosaur was probably an ambush predator, lying in wait to pounce on prey. The fossil of a young one has been found with tooth marks made by the giant crocodile Deinosuchus. The bone had healed, so the dinosaur had escaped!

PERIOD	Late Cretaceous
FAMILY	Tyrannosauridae
DIET	carnivore
LENGTH	33 ft.
WEIGHT	2.2 tons
FINDS	North America

OCTOBER

• October 1st •

TALENKAUEN

One of the earliest iguanodons yet discovered, Talenkauen – the name means "small skull" – lived in what is now Argentina. It had a long neck for its size and, unlike its later relatives, small teeth in the tip of its beak to nip off tasty leaves.

PERIOD	Late Cretaceous
FAMILY	early iguanodont
DIET	herbivore
LENGTH	13 ft.
WEIGHT	145 lb.
FINDS	South America

· October 2nd ·

XIXIANYKUS

These small theropods were efficient dinosaur "roadrunners" with 10-inch-long legs. They could outrun most predators in the forests or across the open plains where they lived. Their short, strong arms ended in massive claws that they used to dig for insects.

PERIOD	Late Cretaceous
FAMILY	Alvarezsauridae
DIET	insectivore
LENGTH	20 in.
WEIGHT	1 lb.
FINDS	Asia

· October 3rd ·

GOYOCEPHALE

Baring the large, sharp teeth at the front of both upper and lower jaws would probably have scared away most attackers. If that did not work, the dinosaur's reinforced skull could have been used to head-butt a predator.

PERIOD	Late Cretaceous
FAMILY	Pachycephalosauridae
DIET	herbivore
LENGTH	7 ft.
WEIGHT	88 lb.
FINDS	Asia

• October 4th •

RATIVATES

The fossils of this ostrich-like theropod were found in what is now Canada. It ran quickly on powerful legs after small animals and insects that it swallowed whole – its beaked mouth had no teeth. It was in turn preyed on by large carnivores such as Gorgosaurus (see p.37).

PERIOD	Late Cretaceous
FAMILY	Ornithomimidae
DIET	omnivore
LENGTH	11 ft.
WEIGHT	210 lb.
FINDS	North America

• October 5th •

CHANGYURAPTOR

This four-winged dromaeosaur had the longest tail feathers yet discovered. It was the size of a turkey with the hind wings on its legs. Fast and nimble, it was good at turning quickly in the air to chase insects and other prey in forests.

PERIOD	Early Cretaceous
FAMILY	Dromaeosauridae
DIET	carnivore
LENGTH	4 ft.
WINGSPAN	21 ft.
WEIGHT	9 lb.
FINDS	Asia

• October 6th •

BARSBOLDIA

A large nose allowed this hadrosaur to make deep, bellowing calls to communicate danger to its herd. Like other duckbills, it used hundreds of continually replaced cheek teeth to grind up its plant food.

PERIOD	Late Cretaceous
FAMILY	Hadrosauridae
DIET	herbivore
LENGTH	33 ft.
WEIGHT	3.9 tons
FINDS	Asia

• October 7th •

FALCARIUS

Slow-moving like a giant ground sloth, this is the earliest known member of its family. Two giant boneyards with hundreds of fossils were uncovered in Utah between 2001 and 2005. The dinosaurs had been killed by a sudden catastrophe of some sort – possibly poisoned by toxic gases from a volcanic spring.

PERIOD	Early Cretaceous
FAMILY	Therizinosauridae
DIET	omnivore
LENGTH	13 ft.
WEIGHT	660 lb.
FINDS	North America

• October 8th •

SIATS

Usually slow-moving, Siats could run fast enough in short spurts to catch sauropod prey. It also scavenged along the coast of the Western Interior Seaway, an ancient sea that stretched from today's Mexico to Canada. To the Native American Ute tribe, a Siat is a mythical monster that ate humans.

PERIOD	Late Cretaceous
FAMILY	Neovenatoridae
DIET	carnivore
LENGTH	43 ft.
WEIGHT	5.5 tons
FINDS	North America

BONE-HEADED DINOSAURS

The pachycephalosaurs get their name from hard heads that were extra bony – some with a braincase more than 8 inches thick. They head-butted each other for females (*see pp.150-151*) or to see off predators.

• October 9th •
COLEPIOCEPHALE

This small biped was one of the earliest in its family. Like other types of pachycephalosaur, the unique shape of its head would have helped others of its own kind recognize it.

PERIOD	Late Cretaceous
FAMILY	Pachycephalosauridae
DIET	herbivore
LENGTH	6 ft.
WEIGHT	71 lb.
FINDS	North America

• October 10th •
HANSSUESIA

The dome on Hanssuesia's thick skull roof was wide and, like others of its kind, would have gotten bigger as the dinosaur grew. Males had a larger dome than females.

PERIOD	Late Cretaceous
FAMILY	Pachycephalosauridae
DIET	herbivore
LENGTH	7 ft.
WEIGHT	110 lb.
FINDS	North America

• October 11th •
STEGOCERAS

With short forelimbs and a large, stiff tail, this dinosaur held its head and neck parallel to the ground when it ran. It was lightly built and its knob-rimmed skull was up to 3 inches thick.

PERIOD	Late Cretaceous
FAMILY	Pachycephalosauridae
DIET	herbivore
LENGTH	7 ft.
WEIGHT	110 lb.
FINDS	North America

• October 12th •
TEXACEPHALE

Named "Texas head" after the US state where its remains were found, this plant eater lived in a herd, foraging for plants and seeds to eat on coastal marshes.

PERIOD	Late Cretaceous
FAMILY	Pachycephalosauridae
DIET	herbivore
LENGTH	7 ft.
WEIGHT	100 lb.
FINDS	North America

• October 13th •
STYGIMOLOCH

One of the largest in the family, this dinosaur had one of the most unusual heads. The bony spikes were up to 4 inches long and may have been used to impress females or to fight off other males. Its name means "horned demon from hell."

PERIOD	Late Cretaceous
FAMILY	Pachycephalosauridae
DIET	herbivore
LENGTH	11 ft.
WEIGHT	175 lb.
FINDS	North America

• October 14th •
PRENOCEPHALE

A wonderfully preserved skull of this herd animal that fed on fruit and leaves was found in Mongolia. Most pachycephalosaurs have been found in North America. Prenocephale had a knobbly ridge of bone running round the edge of its domed head.

PERIOD	Late Cretaceous
FAMILY	Pachycephalosauridae
DIET	herbivore
LENGTH	6 ft.
WEIGHT	110 lb.
FINDS	Asia

• October 15th •

ANTARCTOPELTA

This is the first armored dinosaur known to have lived on the humid and densely forested landmass that is today's icy Antarctica. It was well protected by the armor on its back and sides, and its fearsome spikes.

PERIOD	Late Cretaceous
FAMILY	Nodosauridae
DIET	herbivore
LENGTH	13 ft.
WEIGHT	1.3 tons
FINDS	Antarctica

• October 16th •

NASUTOCERATOPS

With a big nose and the longest horns in its family, this dinosaur roamed in a herd to find the plants it ate. It lived in the swamplands of an island continent called Laramidia, now part of North America.

PERIOD	Late Cretaceous
FAMILY	Ceratopsidae
DIET	herbivore
LENGTH	16 ft.
WEIGHT	2.8 tons
FINDS	North America

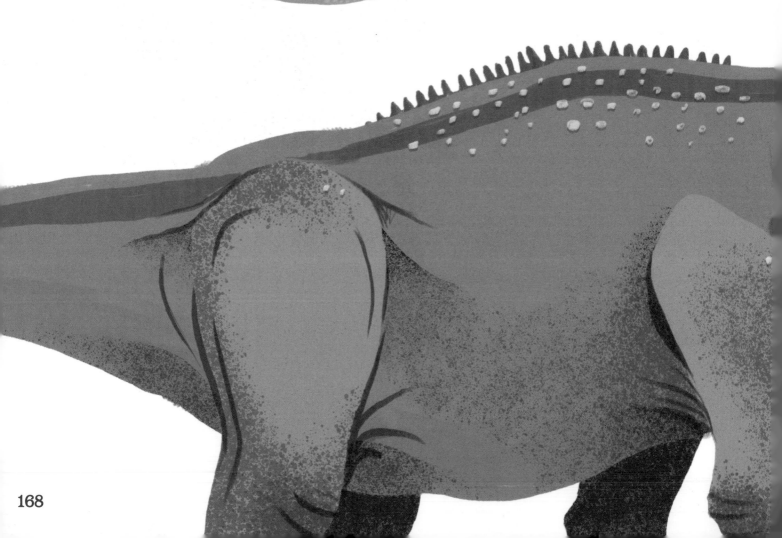

• October 17th •
HEXINLUSAURUS

This dinosaur moved in a large herd through the lush forests and beside the wide rivers of its habitat in what is now China. It fed on low-growing plants with its short, beak-like muzzle, chewing them with the teeth in its muscular cheeks. It was small but fast, able to run quickly away if a predator such as Gasosaurus (see p.137) was on the prowl.

PERIOD	Middle Jurassic
FAMILY	Early ornithischian
DIET	herbivore
LENGTH	6 ft.
WEIGHT	44 lb.
FINDS	Asia

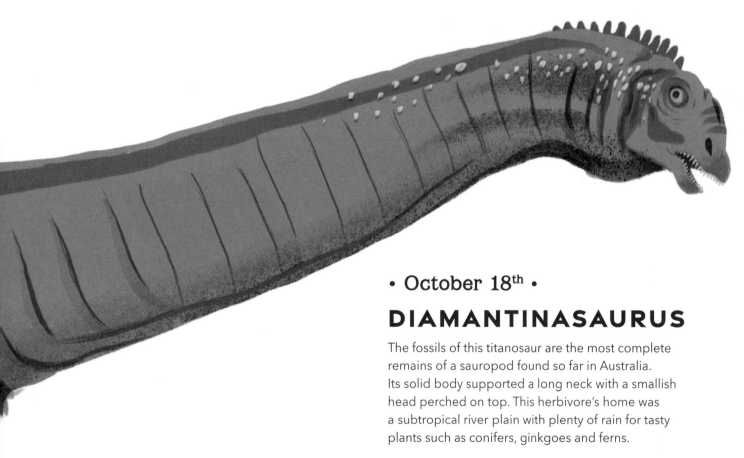

• October 18th •
DIAMANTINASAURUS

The fossils of this titanosaur are the most complete remains of a sauropod found so far in Australia. Its solid body supported a long neck with a smallish head perched on top. This herbivore's home was a subtropical river plain with plenty of rain for tasty plants such as conifers, ginkgoes and ferns.

PERIOD	Late Cretaceous
FAMILY	Titanosauridae
DIET	herbivore
LENGTH	52 ft.
WEIGHT	22 tons
FINDS	Australia

• October 19th •

JIANGXISAURUS

This oviraptor was a small, feathered theropod. Its skull was short and narrow with crushing, toothless jaws to demolish the mollusks and tough plants that it ate. It had three-fingered grasping hands with long, curved claws.

PERIOD	Late Cretaceous
FAMILY	Oviraptoridae
DIET	omnivore
LENGTH	7 ft.
WEIGHT	130 lb.
FINDS	Asia

• October 20th •

SCIPIONYX

A small but swift hunter, Scipionyx would eat anything it could catch, including insects, lizards and fish. It also needed speed to avoid crocodile-like reptiles and larger dinosaurs.

PERIOD	Early Cretaceous
FAMILY	Compsognathidae
DIET	carnivore
LENGTH	7 ft.
WEIGHT	130 lb.
FINDS	Europe

• October 21ˢᵗ •

NANYANGOSAURUS

Living in a herd, this early duck-billed dinosaur used its beaked snout to browse on low-lying plants, although it could stand on two legs to reach tastier leaves. Thousands of Nanyangosaurus fossil eggs have been found on one site in today's central China.

PERIOD	Early Cretaceous
FAMILY	Hadrosauridae
DIET	herbivore
LENGTH	16 ft.
WEIGHT	1.4 tons
FINDS	Asia

• October 22ⁿᵈ •

GEMINIRAPTOR

A lightweight relative of raptors such as Deinonychus (see pp.118–119), this troodont had a small head that contained a relatively big brain. Its large eyes and sharp claws helped it spot and trap small prey. It had an unusual hollow jawbone that some scientists think it may have used to make sounds.

PERIOD	Early Cretaceous
FAMILY	Troodontidae
DIET	carnivore
LENGTH	5 ft.
WEIGHT	20 lb.
FINDS	North America

• October 23rd •

WULONG

With a long tail twice its length and a narrow face filled with sharp teeth, it is no surprise that this small raven-sized raptor's name means "dancing dragon." Wulong could not fly but glided from tree to tree in the forests where it lived.

PERIOD	Early Cretaceous
FAMILY	Dromaeosauridae
DIET	carnivore
LENGTH	2 ft.
WEIGHT	13 lb.
FINDS	Asia

• October 24th •

PISANOSAURUS

Fossils of this small herbivore were discovered in what is now Argentina. This is one of the oldest plant-eating dinosaurs found so far. Unlike its later relatives, it was lightly built and walked on two legs. It had closely packed teeth to feed on soft, low-growing plants.

PERIOD	Late Triassic
FAMILY	Pisanosauridae
DIET	herbivore
LENGTH	3 ft.
WEIGHT	20 lb.
FINDS	South America

• October 25th •

ACANTHOPHOLIS

This low-slung, slow-moving quadruped was well protected from fierce predators by the hard armored plates on its head, neck, back and tail. However, it did not have a club on the end of its tail like other ankylosaurs. It foraged for plants in the forests of what is today's western Europe.

PERIOD	Late Cretaceous
FAMILY	Nodosauridae
DIET	herbivore
LENGTH	18 ft.
WEIGHT	3 tons
FINDS	Europe

TYRANNOSAUR TYRANTS

Early members of this fascinating dinosaur family were small carnivores of the late Jurassic, but their descendants grew to dominate the Cretaceous with some of the fiercest creatures that have ever lived.

• October 26th •
ALIORAMUS

A very good sense of smell would have helped this "terror lizard" to locate sauropods or hadrosaurs to eat or scavenge on the floodplains where it lived.

PERIOD	Late Cretaceous
FAMILY	Tyrannosauridae
DIET	carnivore
LENGTH	20 ft.
WEIGHT	815 lb.
FINDS	Asia

• October 27th •
DASPLETOSAURUS

This terrifying, heavy-boned dinosaur had the largest teeth of any tyrannosaur. They were curved and saw-edged to slice through bone. Its name means "frightful lizard."

PERIOD	Late Cretaceous
FAMILY	Tyrannosauridae
DIET	carnivore
LENGTH	30 ft.
WEIGHT	3.3 tons
FINDS	North America

• October 28th •
TERATOPHONEUS

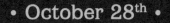

With its strong jaw muscles, this tyrannosaur would have had an impressive bite and it needed it to see off the giant swamp alligators where it lived.

PERIOD	Late Cretaceous
FAMILY	Tyrannosauridae
DIET	carnivore
LENGTH	21 ft.
WEIGHT	1.1 tons
FINDS	North America

• October 29th •
NANUQSAURUS

The remains of this meat eater were found in what is now Alaska, the only tyrannosaur that has been found that far north. At the time, the area would have had a warm climate.

PERIOD	Late Cretaceous
FAMILY	Tyrannosauridae
DIET	carnivore
LENGTH	23 ft.
WEIGHT	1.4 tons
FINDS	North America

• October 30th •
AVIATYRANNIS

One of the oldest members of the family, this nimble biped was small and more likely to be prey than predator. This would not have stopped it hunting for small animals to eat.

PERIOD	Late Jurassic
FAMILY	Tyrannosauridae
DIET	carnivore
LENGTH	4 ft.
WEIGHT	11 lb.
FINDS	Europe

• October 31st •
TARBOSAURUS

The largest predator in what is now Mongolia, this dinosaur was much like Tyrannosaurus (see p.113), but it had a longer skull and was not so heavily built.

PERIOD	Late Cretaceous
FAMILY	Tyrannosauridae
DIET	carnivore
LENGTH	39 ft.
WEIGHT	6.1 tons
FINDS	Asia

NOVEMBER

• November 1st •
TSAAGAN

This medium-sized feathered dromaeosaur, or "running lizard," flourished in an area of the world now called Mongolia. It was an efficient hunter with very strong jaws. A pack of the fierce Tsaagan worked together to trap and kill smaller dinosaurs such as Shuvuuia (*above and p.156*), to add to their regular diet of lizards and small mammals.

PERIOD	Late Cretaceous
FAMILY	Dromaeosauridae
DIET	carnivore
LENGTH	7 ft.
WEIGHT	44 lb.
FINDS	Asia

• November 2nd •

DAEMONOSAURUS

This dog-sized theropod had unusually sharp and pointed teeth to hold struggling prey. It would have hidden in the undergrowth to ambush small animals.

PERIOD	Late Triassic
FAMILY	early theropod
DIET	carnivore
LENGTH	7 ft.
WEIGHT	55 lb.
FINDS	North America

• November 3rd •

HUAXIAGNATHUS

An active hunter of small mammals and reptiles, this carnivore would also have captured and killed small dinosaurs such as Sinocalliopteryx (see p.36).

PERIOD	Early Cretaceous
FAMILY	Compsognathidae
DIET	carnivore
LENGTH	6 ft.
WEIGHT	44 lb.
FINDS	Asia

• November 4th •

ARRHINOCERATOPS

Well protected by its long forward-curving horns, this plant eater had only a short nose horn. It foraged for plants to eat as it traveled in a herd.

PERIOD	Late Cretaceous
FAMILY	Ceratopsidae
DIET	herbivore
LENGTH	20 ft.
WEIGHT	2.2 tons
FINDS	North America

• November 5th •

CRYOLOPHOSAURUS

Living in a warm forested area along the coastline of what is now icy Antarctica, this carnivore had no trouble finding plenty of its favorite plant-eating prey to eat. This was the largest known theropod that lived in this area of the world at the time. It had a furrowed crest rising vertically from the top of its head that was used for display to attract a mate or see off a rival.

PERIOD	Early Jurassic
FAMILY	early theropod
DIET	carnivore
LENGTH	8 ft.
WEIGHT	1,450 lb.
FINDS	Antarctica

PEACEFUL PLANT EATERS

The large, slow iguanodonts could move in their herds on either two or four legs. They had spiked thumbs that they may have used as weapons, but which would also have helped them pick up fruits and other plants to eat.

• November 6th •
CALLOVOSAURUS

This is the earliest iguanodont to have been found so far. It grazed alongside other plant-eating dinosaurs, including sauropods and stegosaurs.

PERIOD	Middle Jurassic
FAMILY	Dryosauridae
DIET	herbivore
LENGTH	9 ft.
WEIGHT	275 lb.
FINDS	Europe

• November 7th •
IGUANACOLOSSUS

This heavily built dinosaur traveled in a herd with its young, but may have been more independent when fully grown. It was probably preyed on by Utahraptor (see p.197).

PERIOD	Early Cretaceous
FAMILY	Iguanodontidae
DIET	herbivore
LENGTH	33 ft.
WEIGHT	2.2 tons
FINDS	North America

• November 8th •
ATLASCOPCOSAURUS

This small plant eater may have been prey to the large predator Australovenator. The area in today's Victoria, Australia, where it lived is now called Dinosaur Cove.

PERIOD	Early Cretaceous
FAMILY	Hypsilophodontidae
DIET	herbivore
LENGTH	10 ft.
WEIGHT	275 lb.
FINDS	Australia

• November 9th •
MANTELLISAURUS

Trackways have been found showing that this large iguanodont traveled in a family group. It had short forelimbs, so on all four legs it would probably only have walked slowly or stood still.

PERIOD	Early Cretaceous
FAMILY	Iguanodontidae
DIET	herbivore
LENGTH	23 ft.
WEIGHT	1,750 lb.
FINDS	Europe

• November 10th •
MOCHLODON

Weighing around the same as an adult male lion, this iguanodont lived in woodlands. It is thought that it was small because it lived on an island where food was limited, so it had to adapt over time.

PERIOD	Late Cretaceous
FAMILY	Rhabdodontidae
DIET	herbivore
LENGTH	10 ft.
WEIGHT	550 lb.
FINDS	Europe

• November 11th •
CAMPTOSAURUS

With a long, beak-ended snout full of teeth, this early plant eater grazed on all fours, but could balance and walk on two legs to reach leaves higher up, possibly using its tail as a prop.

PERIOD	Late Jurassic
FAMILY	Camptosauridae
DIET	herbivore
LENGTH	23 ft.
WEIGHT	1.1 tons
FINDS	North America

• November 12th •
AVIMIMUS

With very long legs for running after prey or away from predators, this fast-moving "bird mimic" lived in large flocks in the wetlands that existed at that time. It was ostrich-like and could not fly, but it was able to fold its arms close to its body like birds fold their wings today.

PERIOD	Late Cretaceous
FAMILY	Avimimidae
DIET	omnivore
LENGTH	5 ft.
WEIGHT	33 lb.
FINDS	Asia

• November 13th •
BONAPARTENYKUS

The largest known member of its family, this small theropod had short, stout arms with claws to dig in termite mounds and ant hills. Fossils of this dinosaur have been found in today's Argentina, together with very unusual fossilized eggs.

PERIOD	Late Cretaceous
FAMILY	Alvarezsauridae
DIET	insectivore
LENGTH	8 ft.
WEIGHT	100 lb.
FINDS	South America

• November 14th •

MINMI

In the Early Cretaceous, most of eastern Australia was covered by a shallow sea, and Minmi lived on its coastal plains. It was a small plant eater, covered in bony armor, that lived in a herd.

PERIOD	Early Cretaceous
FAMILY	Ankylosauridae
DIET	herbivore
LENGTH	10 ft.
WEIGHT	815 lb.
FINDS	Australia

• November 15th •

MICROVENATOR

This turkey-sized "small hunter" had a large head and beak, as well as toothless jaws. It would have found a variety of food, swallowing whole the mammals, reptiles and insects that it hunted, and also eating plants. Like other oviraptors, it would have used its feathers for display.

PERIOD	Early Cretaceous
FAMILY	Caenagnathidae
DIET	omnivore
LENGTH	10 ft.
WEIGHT	110 lb.
FINDS	North America

• November 16th •

ICHTHYOSAURUS

Small and dolphin-like, Ichthyosaurus was a powerful swimmer, reaching speeds of up to 25 mph. A pod of this swimming reptile needed to be quick to escape the jaws of a threatening Temnodontosaurus (*center left and p.123*)! Ichthyosaurus in turn hunted fish, ammonites, octopus and squid, which it caught with its long, narrow snout full of sharp teeth.

PERIOD	Early Jurassic
FAMILY	Ichthyosauridae
DIET	piscivore
LENGTH	7 ft.
WEIGHT	200 lb.
FINDS	Europe, North America, Greenland

GIANT SAUROPODS

Remains of these huge, four-footed plant eaters have been found on every continent except Antarctica. These were the true dinosaur giants, shaking the earth wherever they trod, the largest animals to have ever lived on land.

• November 17th •
SALTASAURUS

Although still massive, this was one of the smaller titanosaurs, with a shorter neck and limbs. It laid very large eggs with shells that were 0.2 inches thick – the thickest known so far.

PERIOD	Late Cretaceous
FAMILY	Saltasauridae
DIET	herbivore
LENGTH	39 ft.
WEIGHT	11 tons
FINDS	South America

• November 18th •
HYPSELOSAURUS

Living in woodlands in what is now southern France, this titanosaur had unusually thick legs. It may have been one of the last of the titanosaurs.

PERIOD	Late Cretaceous
FAMILY	Titanosauridae
DIET	herbivore
LENGTH	39 ft.
WEIGHT	11 tons
FINDS	Europe

• November 19th •
ALAMOSAURUS

This is the only titanosaur known from North America. Eating up to 33 pounds a day, it ranged over vast areas stripping vegetation as it went.

PERIOD	Late Cretaceous
FAMILY	Saltasauridae
DIET	herbivore
LENGTH	79 ft.
WEIGHT	26.5 tons
FINDS	North America

• November 20th •
XINJIANGTITAN

This plant eater could browse on very high trees. It had an immensely long neck, measuring about half its length, with some of the vertebrae measuring 3 feet or more!

PERIOD	Middle Jurassic
FAMILY	Mamenchisauridae
DIET	herbivore
LENGTH	100 ft.
WEIGHT	33.1 tons
FINDS	Asia

• November 21st •
MALAWISAURUS

Bony growths on the back of this early, heavy-set titanosaur's neck and along its back acted as armor to protect it from attacks from predators.

PERIOD	Early Cretaceous
FAMILY	Titanosauridae
DIET	herbivore
LENGTH	52 ft.
WEIGHT	26.5 tons
FINDS	Africa

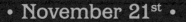

• November 22nd •
ARGYROSAURUS

This "silver lizard" was one of the first South American titanosaurs to be named. It traveled across what is now Argentina in a herd, using its very long neck to reach the tops of the trees for tasty leaves. This was a big dinosaur - its tibia, or upper leg bone, was more than 6 feet in length!

PERIOD	Late Cretaceous
FAMILY	Titanosauridae
DIET	herbivore
LENGTH	92 ft.
WEIGHT	28.7 tons
FINDS	South America

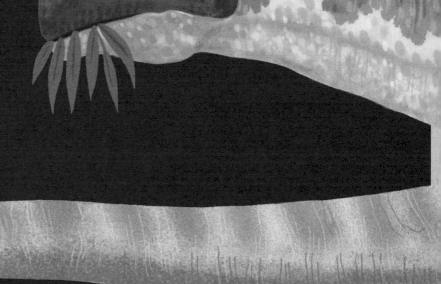

• November 23rd •
SALTRIOVENATOR

A very early member of its family, this predator was the largest and most agile carnivore of its time. It roamed coastal forests in what is now Italy, feeding on any small to medium-sized animals it found. It may have even waded into the sea after fish or sharks swimming in the shallows.

PERIOD	Early Jurassic
FAMILY	Ceratosauridae
DIET	carnivore
LENGTH	33 ft.
WEIGHT	1.1 tons
FINDS	Europe

• November 24th •
TRICERATOPS

This plant eater had an enormous skull and a backward-leaning frill that flushed red with blood to attract females or warn of danger. Behind its hard beak were jaws housing batteries of shearing teeth that, when they had been worn down by tough vegetation, were replaced by yet more. A herd of Triceratops was capable of defending their young against an attack by Tyrannosaurus (see p.113).

PERIOD	Late Cretaceous
FAMILY	Ceratopsidae
DIET	herbivore
LENGTH	30 ft.
WEIGHT	7.2 tons
FINDS	North America

• November 25th •

SAUROPOSEIDON

This plant eater is possibly the tallest sauropod, with a neck up to 39 feet in length. The large vertebrae inside were honeycombed with tiny air pockets, making the neck lighter in weight and so easier to lift. When fossils were first found in Oklahoma, USA, in 1994, their large size led people to classify them as petrified tree trunks and not as parts of a dinosaur!

PERIOD	Early Cretaceous
FAMILY	Brachiosauridae
DIET	herbivore
LENGTH	105 ft.
WEIGHT	66.1 tons
FINDS	North America

• November 26th •

HYPSELOSPINUS

This tall-spined iguanodon roamed what is now southern England looking for the low-growing plants it liked to eat. Like other members of its family, it lived in a herd, which gave it some protection from predators such as Eotyrannus.

PERIOD	Early Cretaceous
FAMILY	Iguanodontidae
DIET	herbivore
LENGTH	20 ft.
WEIGHT	1,750 lb.
FINDS	Europe

• November 27th •

ASHDOWN MANIRAPTORAN

Among the smallest of the known dinosaurs, this bird-like but flightless theropod hunted animals and snapped at insects, but also ate leaves and fruit. It had a long neck and slim hind legs much like the wading birds that exist today.

PERIOD	Early Cretaceous
FAMILY	Coelurosauridae
DIET	omnivore
LENGTH	1 ft.
WEIGHT	7 oz.
FINDS	Europe

• November 28th •
NODOCEPHALOSAURUS

Heavily armored, this ankylosaur also had a large tail club to defend itself against predators. The shape of its head is very similar to those of the ankylosaurs of Asia. This may be evidence that in the time of the dinosaurs there was a land bridge between what is now North America and Asia.

PERIOD	Late Cretaceous
FAMILY	Ankylosauridae
DIET	herbivore
LENGTH	15 ft.
WEIGHT	1.7 tons
FINDS	North America

• November 29th •
AURORACERATOPS

An early relative of Triceratops (see p. 188), this biped did not have the armor of the later dinosaur. It used its short snout and fang-like teeth to dig in the ground and pull out plants to eat.

PERIOD	Early Cretaceous
FAMILY	Ceratopsidae
DIET	herbivore
LENGTH	7 ft.
WEIGHT	220 lb.
FINDS	Asia

• November 30th •
SHANSHANOSAURUS

So far, this is the smallest known tyrannosaur. It was active during twilight, when the light is dim, so had large eyes in its long skull. These helped it to spot swift-moving lizards and scuttling mammals in the forest undergrowth.

PERIOD	Late Cretaceous
FAMILY	Tyrannosauridae
DIET	carnivore
LENGTH	10 ft.
WEIGHT	200 lb.
FINDS	Asia

DECEMBER

• December 1st •
FOSTORIA

The remains of a whole herd of Fostoria were found in a remote Australian opal mine. The fossils had become "opalized" when they turned to stone, and light bouncing off them makes them very colorful. This plant eater lived on a floodplain that was rich in vegetation, with rivers flowing into the inland Eromanga Sea.

PERIOD	Late Cretaceous
FAMILY	Iguanodontidae
DIET	herbivore
LENGTH	16 ft.
WEIGHT	2.8 tons
FINDS	Australia

• December 2nd •

XUWULONG

Named "Xuwu dragon" by the paleontologists who found it, this very early hadrosaur lived in what is now northwest China. It had a shorter skull than other members of the family and its lower jaw was more of a V shape, which may mean that it ate unusual plant food or had a different way of eating. Like other hadrosaurs, it lived and moved in a herd.

PERIOD	Early Cretaceous
FAMILY	Hadrosauridae
DIET	herbivore
LENGTH	7 ft.
WEIGHT	1.1 tons
FINDS	Asia

• December 3rd •

AGUJACERATOPS

This small horned dinosaur lived in the wetlands near the coast of the Western Interior Seaway that existed in what is now central North America. It had a wide frill and long brow horns that it could use when attacked by Deinosuchus, a giant crocodile.

PERIOD	Late Cretaceous
FAMILY	Ceratopsidae
DIET	herbivore
LENGTH	16 ft.
WEIGHT	1.7 tons
FINDS	North America

• December 4th •

BUITRERAPTOR

Lightweight and nimble, this turkey-sized "vulture-raptor" hunted small lizards and mammals in the rocky landscape of what is today's Argentina. Its long, three-fingered hands did not have the sickle claw typical of later dromaeosaurs. It was probably preyed on by large meat eaters such as Mapusaurus (*see p.88*) and Giganotosaurus (*see p.197*).

PERIOD	Late Cretaceous
FAMILY	Dromaeosauridae
DIET	carnivore
LENGTH	4 ft.
WEIGHT	22 lb.
FINDS	South America

• December 5th •

AKAINACEPHALUS

With the large bony club at the end of its tail, Akainacephalus was able to protect itself against the fierce tyrannosaur Bistahieversor (*see p.79*) in the forests where it lived. The bony armor covering this North American ankylosaur's snout and head is very similar to that of Asian ankylosaurs such as Tarchia (*see p.75*). This may be evidence that the Asian dinosaurs migrated to North America across a land bridge in the Late Cretaceous.

PERIOD	Late Cretaceous
FAMILY	Ankylosauridae
DIET	herbivore
LENGTH	16 ft.
WEIGHT	1.1 tons
FINDS	North America

BIPED HUNTERS

There were some very efficient hunters among the carnivores, and moving on two legs gave them the speed they needed. It also freed their forelimbs to grab or injure prey, although in many later dinosaurs these got smaller.

• December 6th •
BARYONYX

Wading in rivers and river deltas after its fish prey, this crocodile-snouted dinosaur also caught smaller dinosaurs on land.

PERIOD	Early Cretaceous
FAMILY	Spinosauridae
DIET	piscivore
LENGTH	33 ft.
WEIGHT	1.9 tons
FINDS	Europe

• December 7th •
SIAMRAPTOR

Running down iguanodonts and other plant eaters, this large predator was merciless with its blade-like teeth. Its remains were found in today's Thailand.

PERIOD	Early Cretaceous
FAMILY	Carcharodontosauridae
DIET	carnivore
LENGTH	30 ft.
WEIGHT	3.3 tons
FINDS	Asia

• December 8th •
DILOPHOSAURUS

This crested carnivore probably used scent to find its prey. It hunted in packs and could have reached speeds of up to 31 mph in pursuit.

PERIOD	Early Jurassic
FAMILY	Dilophosauridae
DIET	carnivore
LENGTH	23 ft.
WEIGHT	1 ton
FINDS	North America

• December 9th •
UTAHRAPTOR

When it ran, Utahraptor angled its 9-inch-long sickle claws away from the ground. When a claw had clamped down on prey, sharp teeth finished it off.

PERIOD	Early Cretaceous
FAMILY	Dromaeosauridae
DIET	carnivore
LENGTH	25 ft.
WEIGHT	1 ton
FINDS	North America

• December 10th •
ELAPHROSAURUS

Its light weight meant that this carnivore would have relied on speed to catch small animals, but it also needed to move fast to escape predators.

PERIOD	Late Jurassic
FAMILY	Noasauridae
DIET	carnivore
LENGTH	20 ft.
WEIGHT	550 lb.
FINDS	Africa

• December 11th •
GIGANOTOSAURUS

This "giant southern lizard" was one of the largest of the meat eaters and could move its powerful legs up to 31 mph. Its huge skull housed teeth up to 8 inches long.

PERIOD	Late Cretaceous
FAMILY	Carcharodontosauridae
DIET	carnivore
LENGTH	43 ft.
WEIGHT	14 tons
FINDS	South America

• December 12th •
MEI LONG

The first fossil that was found of this small, bird-like troodont had its head tucked under its wing, so it has a Chinese name meaning "soundly sleeping dragon." This position is like that of many modern birds. Mei Long caught small mammals, lizards and hatchling dinosaurs, but also ate leaves.

PERIOD	Early Cretaceous
FAMILY	Troodontidae
DIET	omnivore
LENGTH	1 ft.
WEIGHT	6 lb.
FINDS	Asia

• December 13th •
ALBALOPHOSAURUS

In the underbrush of the woodlands where it lived, this early member of its family needed to be a speedy runner to avoid predators. Its name means "white crested lizard," but this does not describe its appearance – it refers to the white-topped Japanese mountain near which fossils were found.

PERIOD	Early Cretaceous
FAMILY	early ceratopsian
DIET	herbivore
LENGTH	7 ft.
WEIGHT	88 lb.
FINDS	Asia

• December 14th •
PATAGONYKUS

This feathered theropod was adapted for its specialized diet, feeding on insects with a long, tube-shaped snout that housed very small teeth. It had long legs and short arms with a single clawed finger on each hand. Using its claws powered by large breast and arm muscles, it dug into and tore open ant and termite nests, sucking up the contents.

PERIOD	Late Cretaceous
FAMILY	Alvarezsauridae
DIET	insectivore
LENGTH	8 ft.
WEIGHT	66 lb.
FINDS	South America

• December 15th •
COELOPHYSIS

A little larger than a turkey, this was one of the earliest dinosaurs. It lived in a large pack and was a quick hunter of fish, small dinosaurs, lizards and insects. It was long and slender, with large eyes that it needed to spot prey and its crocodile-like predators.

PERIOD	Late Triassic
FAMILY	Coelophysidae
DIET	carnivore
LENGTH	7 ft.
WEIGHT	55 lb.
FINDS	North America

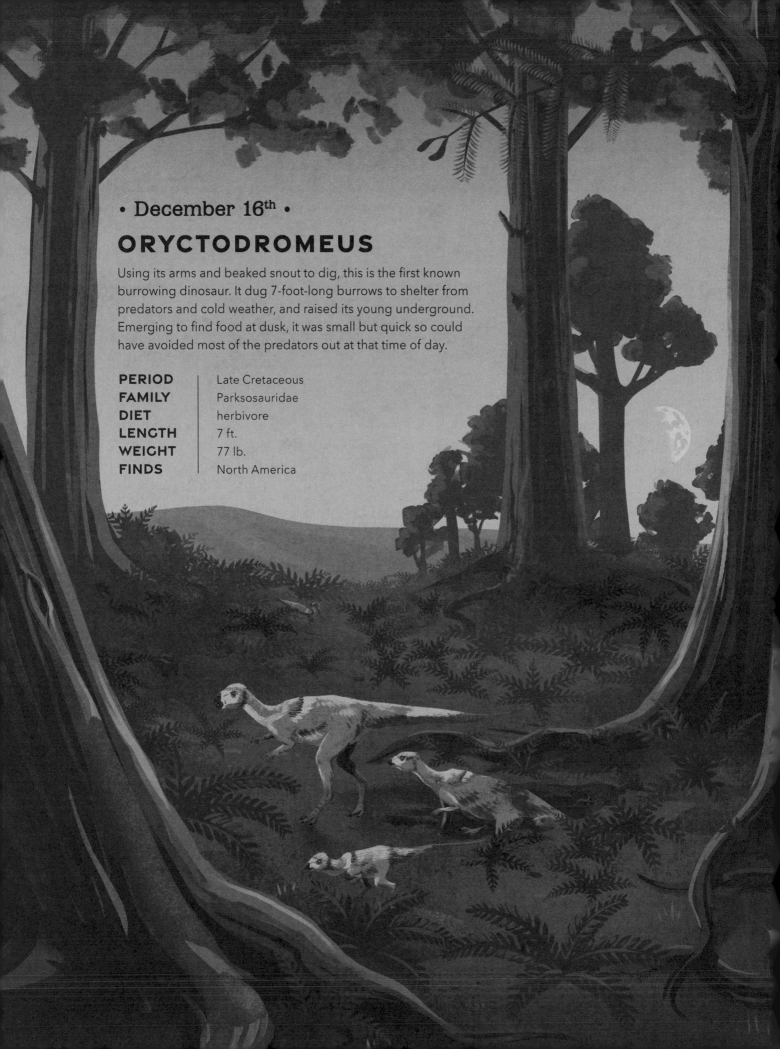

• December 16th •

ORYCTODROMEUS

Using its arms and beaked snout to dig, this is the first known burrowing dinosaur. It dug 7-foot-long burrows to shelter from predators and cold weather, and raised its young underground. Emerging to find food at dusk, it was small but quick so could have avoided most of the predators out at that time of day.

PERIOD	Late Cretaceous
FAMILY	Parksosauridae
DIET	herbivore
LENGTH	7 ft.
WEIGHT	77 lb.
FINDS	North America

• December 17th •

OMEISAURUS

With more vertebrae – it had 17 of them – that were also longer and larger than those of many other plant eaters, this dinosaur had a remarkably long neck. However, the neck was also lightweight, allowing Omeisaurus to reach its tiny head right up to the very tops of trees, although it also browsed in herds on low-growing seed ferns and shrubs.

PERIOD	Late Jurassic
FAMILY	Mamenchisauridae
DIET	herbivore
LENGTH	66 ft.
WEIGHT	17.6 tons
FINDS	Asia

• December 18th •
NODOSAURUS

Remains of this armored dinosaur were found by miners in Canada and were the best preserved fossils of its family ever found. It lived in conifer forests and meadows near an inland sea, and was protected by its armored plates and spikes.

PERIOD	Late Cretaceous
FAMILY	Nodosauridae
DIET	herbivore
LENGTH	18 ft.
WEIGHT	1.4 tons
FINDS	North America

• December 19th •
PROA

This plant eater's lower jaw reminded its finders of the bow of a boat. As its remains were discovered in a coal mine in Spain, it was named "proa," which means "prow" in Spanish. It moved in a herd across the wetlands browsing on low lying plants.

PERIOD	Early Cretaceous
FAMILY	Iguanodontidae
DIET	herbivore
LENGTH	26 ft.
WEIGHT	1.1 tons
FINDS	Europe

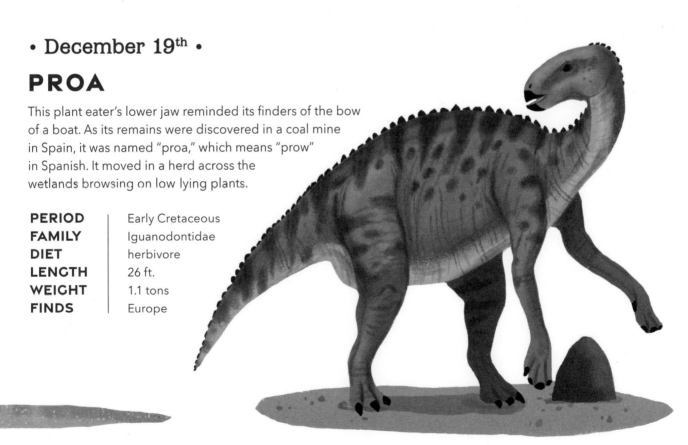

• December 20th •

AUCASAURUS

Unusually, an almost complete skeleton of this dinosaur has been found in Argentina. As well as telling us more about how the dinosaur lived, it also gave scientists an idea of how its skin looked. It had long powerful legs and strong jaws, but very short arms and no claws. It was a pack hunter, feeding on sauropods such as Saltasaurus (see p. 186).

PERIOD	Late Cretaceous
FAMILY	Abelisauridae
DIET	carnivore
LENGTH	20 ft.
WEIGHT	1,550 lb.
FINDS	South America

• December 21st •

AJKACERATOPS

This small plant eater had a beaked mouth like a parrot. It lived in woodlands, grazing in herds on the low-lying bushes of a floodplain. In the Late Cretaceous, Europe was really a series of pieces of land surrounded by sea rather than a continent.

PERIOD	Late Cretaceous
FAMILY	Bagaceratopsidae
DIET	herbivore
LENGTH	3 ft.
WEIGHT	44 lb.
FINDS	Europe

• December 22nd •
YI

With wings like a bat, this pigeon-sized insectivore glided through the trees snapping up flies, dragonflies and moths. What are unique in this dinosaur are the rod-like bones extending from its wrists and connecting to its "wings," like those in today's flying squirrels.

PERIOD	Late Jurassic
FAMILY	Scansoriopterygidae
DIET	insectivore
LENGTH	2 ft.
WEIGHT	13 oz.
FINDS	Asia

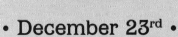

• December 23rd •
LYCORHINUS

One of the first of the known Jurassic dinosaurs to be found in Africa, Lycorhinus ate small animals and snipped vegetation, but spent much of its time hiding from the agile theropods that shared its habitat.

PERIOD	Early Jurassic
FAMILY	Heterodontosauridae
DIET	omnivore
LENGTH	4 ft.
WEIGHT	15 lb.
FINDS	Africa

• December 24th •
FRUITADENS

This small dinosaur was a fast runner, able to escape the jaws of giant predators such as Torvosaurus (see p.60). It lived in woodlands and ate small animals and bugs as well as plants.

PERIOD	Late Jurassic
FAMILY	Heterodontosauridae
DIET	omnivore
LENGTH	2 ft.
WEIGHT	2 lb.
FINDS	North America

• December 25th •

NEMEGTOSAURUS

Raising its long face on its long neck to the tops of trees, this titanosaur (see pp.186-187) used its pencil-shaped teeth to clip off the leaves and tasty flowers. At the time, there were plenty of flowering plants, ferns and conifers to eat in the area of what is today's Gobi desert that stretches from Mongolia to China.

PERIOD	Late Cretaceous
FAMILY	Nemegtosauridae
DIET	herbivore
LENGTH	49 ft.
WEIGHT	19.8 tons
FINDS	Asia

• December 26th •

HYPACROSAURUS

This large duckbill, like others in its family, was a popular prey animal for tyrannosaurs and troodons. Its crest was hollow, to amplify warning calls to its herd in the forests where it lived. It laid up to 20 eggs at a time in large nests, which were often raided by the same predators.

PERIOD	Late Cretaceous
FAMILY	Hadrosauridae
DIET	herbivore
LENGTH	30 ft.
WEIGHT	4.4 tons
FINDS	North America

• December 27th •
RHOETOSAURUS

One of the largest and most complete dinosaur fossils discovered in Australia, this early sauropod is also one of the oldest. It traveled in a family group at a gentle walking speed of up to 9 mph through conifer forests near rivers. It browsed on the conifers as well as cycads and tree ferns in what was a wet, humid and warm climate.

PERIOD	Middle Jurassic
FAMILY	early sauropod
DIET	herbivore
LENGTH	49 ft.
WEIGHT	9.9 tons
FINDS	Australia

• December 28th •
KOSMOCERATOPS

This dinosaur's name means "ornamented horned face." The horns were used for defense or possibly to attract females, and this plant eater would have certainly done either of those very successfully! It had 15 horns in all, one over the nose, two over the eyes, two on the cheeks and ten across the back of its frill.

PERIOD	Late Cretaceous
FAMILY	Ceratopsidae
DIET	herbivore
LENGTH	16 ft.
WEIGHT	2.8 tons
FINDS	North America

• December 29th •

ZANABAZAR

One of the largest troodonts, this lightweight dinobird lived in woodlands in what is now Mongolia, central Asia. This was a clever and efficient predator, running quickly on its long hind legs to chase down smaller dinosaurs and mammals and to snap at insects.

PERIOD	Late Cretaceous
FAMILY	Troodontidae
DIET	carnivore
LENGTH	8 ft.
WEIGHT	55 lb.
FINDS	Asia

• December 30th •

MAJUNGASAURUS

With more teeth than any other abelisaurid – 17 in each of the lower and upper jaws – this predator hunted sauropods such as Rapetosaurus (see p.49) as well as smaller members of its own family.

PERIOD	Late Cretaceous
FAMILY	Abelisauridae
DIET	carnivore
LENGTH	26 ft.
WEIGHT	1.7 tons
FINDS	Africa

• December 31st •

RUYANGOSAURUS

A herd of these enormous titanosaurs fed together as they moved through the floodplains and forests of what is today's China. They would have been a magnificent sight, their long necks swaying as they reached for and snipped off the leaves and branches of the trees and shrubs at all levels.

PERIOD	Early Cretaceous
FAMILY	Titanosauridae
DIET	herbivore
LENGTH	82 ft.
WEIGHT	37.5 tons
FINDS	Asia

THE END OF THE DINOSAURS

About 66 million years ago, a massive asteroid or comet hit Earth off the coast of today's Mexico, leaving a crater 93 miles across and 12 miles deep. The collision threw toxic gases and dust into the air that circulated around the planet, blocking out the sunlight. Together with increased volcanic activity, this event caused a global greenhouse effect and climate change as temperatures plummeted. Plants cannot grow without sunlight, so plant eaters starved to death, leaving no prey for the meat eaters to hunt. Sadly, almost all of the dinosaurs became extinct, together with nearly three-quarters of all animal and plant species on Earth at the time.

THE SURVIVORS

Amazingly, some animals survived this disastrous event. Small animals including mammals, snakes, frogs and lizards found a way to live on despite the impact and its aftermath. And crocodiles, sharks, rays and turtles have even remained largely unchanged through the millennia.

The only dinosaurs to survive were a few bird-like species, possibly because they could find seeds despite the plants dying. Today, those relatives of the dinosaurs are the birds, of which there are now more than 18,000 species.

Take a look at the next bird you see – you are looking at a dinosaur!

PRONUNCIATION GUIDE

These pages will help you sound out the different syllables of the complicated and sometimes very long dinosaur names.

A

ABRICTOSAURUS
ah-BRICK-tuh-SAWR-us

ACANTHOPHOLIS
a-kan-THO-phol-iss

ACHELOUSAURUS
ah-KEL-oo-SAWR-us

ACHILLOBATOR
ah-KIL-oh-BAH-tor

ACRISTAVUS
AH-kriss-TA-vus

ACROCANTHOSAURUS
ah-kroh-kan-thuh-SAWR-us

ACROTHOLUS
ACK-roe-THO-luss

ADRATIKLIT
A-DRAT-ee-klit

AEPYORNITHOMIMUS
AY-py-OR-ni-thoh-MEEM-us

AFROVENATOR
aff-ROW-VEN-ah-tor

AGUJACERATOPS
A-gui-ha-SER-ah-tops

AKAINACEPHALUS
ah-KANE-ah-sef-ah-luss

AJKACERATOPS
OI-kah-SER-ah-tops

ALAMOSAURUS
ah-la-mow-SAWR-us

ALBALOPHOSAURUS
al-BAL-of-oh-SAWR-us

ALBERTACERATOPS
al-BERT-ah-SER-ah-tops

ALBERTADROMEUS
al-BERT-ah-droe-me-us

ALBERTONYKUS
al-BERT-oh-NY-kus

ALBERTOSAURUS
al-BERT-oh-SAWR-us

ALECTROSAURUS
AH-lek-tro-SAWR-us

ALETOPELTA
AH-let-oh-PEL-tah

ALIORAMUS
AH-lee-oh-RAY-mus

ALLOSAURUS
AL-oh-sawr-us

ALVAREZSAURUS
AL-var-rez-SAWR-us

ALWALKERIA
AL-wah-KEER-ee-a

AMPHICOELIAS
am-fee-SEEL-ee-yus

ANABISETIA
AH-nah-bis-et-ee-ah

ANCHICERATOPS
AN-chee-SER-ah-tops

ANCHIORNIS
AN-chee-or-niss

ANGATURAMA
ANG-ah-tou-RAH-mah

ANKYLOSAURUS
an-KYE-loh-sawr-us

ANSERIMIMUS
AN-SER-ih-MEEM-us

ANTARCTOPELTA
an-TARK-toe-PEL-tah

ANTARCTOSAURUS
an-TARK-toe-SAWR-us

ANZU
an-ZOU

APATORAPTOR
ah-PAT-oh-RAP-tor

APATOSAURUS
ah-PAT-oh-SAWR-us

APPALACHIOSAURUS
ah-pah-LAY-chee-oh-SAWR-us

AQUILOPS
AH-kwil-ops

ARCHAEOPTERYX
AR-kee-OP-ter-ix

ARGENTINOSAURUS
AR-gen-teen-oh-SAWR-us

ARGYROSAURUS
AR-jy-ro-SAWR-us

ARISTOSUCHUS
ah-RIST-oh-SUE-kus

ARRHINOCERATOPS
ah-RINE-oh-SER-ah-tops

ASHDOWN MANIRAPTORAN
ash-down MAN-ih-RAP-tor-an

ATLASCOPCOSAURUS
AT-lass-KOP-ko-SAWR-us

AUCASAURUS
AW-ka-SAWR-us

AURORACERATOPS
aw-ROR-ah-SER-ah-tops

AUSTRORAPTOR
AW-stroh-rap-tor

AUSTROSAURUS
AW-stroh-SAWR-us

AVACERATOPS
ah-va-SER-ah-tops

AVIATYRANNIS
AY-VEE-ah-tie-RAN-nis

AVIMIMUS
AY-vee-MEEM-us

B

BAHARIASAURUS
BAH-hah-ree-ah-SAWR-us

BAMBIRAPTOR
BAM-bee-RAP-tor

BARAPASAURUS
bar-rah-pah-SORE-us

BARSBOLDIA
bars-BOWL-dee-ah

BARYONYX
bah-ree-ON-iks

BECKLESPINAX
BECK-el-spine-aks

BEIPIAOSAURUS
BAY-pyow-SAWR-us

BISTAHIEVERSOR
BIST-ah-he-ee-VERS-or

BONAPARTENYKUS
BONE-ah-PART-en-eye-kus

BOREALOPELTA
BOR-e-al-PEL-tah

BOROGOVIA
BOR-oh-GOH-vee-a

BRACHIOSAURUS
BRAK-ee-oh-SAWR-us

BRACHYCERATOPS
BRAK-ee-SER-ah-tops

BRACHYLOPHOSAURUS
BRAK-ee-lof-oh-SAWR-us

BUITRERAPTOR
BWEE-tre-RAP-tor

BYRONOSAURUS
BYE-ron-oh-SAWR-us

C

CAIHONG
tsai-hong

CALLOVOSAURUS
kal-lo-voh-SAWR-us

CAMARASAURUS
KAM-ah-rah-SAWR-us

CAMPTOSAURUS
KAMP-toe-SAWR-us

CARCHARODONTOSAURUS
kar-KAR-o-DON-toe-SAWR-us

CARNOTAURUS
kar-NOH-TORE-us

CAUDIPTERYX
kaw-DIP-ter-iks

CENTROSAURUS
sen-troh-SAWR-us

CERATONYKUS
seh-RAT-oh-nyk-us

CERATOSAURUS
seh-RAT-oh-SAWR-us

CHANGYURAPTOR
CHANG-yu-RAP-tor

CHARONOSAURUS
CHAR-oh-noh-SAWR-us

CHASMOSAURUS
CHAZ-moh-SAWR-us

CHILESAURUS
CHILL-ee-SAWR-us

CHINDESAURUS
CHIN-de-SAWR-us

CHIROSTENOTES
CHI-roh-STEN-oh-teez

CITIPATI
chi-ti-pah-tih

COELOPHYSIS
SEE-loh-FIE-sis

COLEPIOCEPHALE
kol-EP-eo-sef-ah-ley

COMPSOGNATHUS
KOMP-sog-NATH-us

CONCAVENATOR
KON-kah-ven-AH-tor

CONCHORAPTOR
KON-koh-RAP-tor

CORYTHORAPTOR
KOH-rith-oh-RAP-tor

CORYTHOSAURUS
KOH-rith-oh-SAWR-us

CRYOLOPHOSAURUS
KRY-oh-lof-oh-SAWR-us

D

DACENTRURUS
DAH-sen-TROOR-us

DAEMONOSAURUS
DAY-mon-oh-SAWR-us

DASPLETOSAURUS
das-PLEE-toe-SAWR-us

DEINOCHEIRUS
DIE-noh-KIRE-us

DEINONYCHUS
DIE-non-IKE-us

DELAPPARENTIA
de-LAP-pah-ren-tee-ah

DELTADROMEUS
del-tah-drohm-ee-us

DIABLOCERATOPS
DEE-ab-low-SER-ah-tops

DIAMANTINASAURUS
DEE-ah-MAN-teen-ah-SAWR-us

DICREAOSAURUS
die-KRAY-oh-SAWR-us

DILOPHOSAURUS
die-LOAF-oh-SAWR-us

DIPLODOCUS
DIP-low-DOC-us

DREADNOUGHTUS
DREAD-not-us

DROMAEOSAURUS
DROM-ee-oh-SAWR-us

DROMICEIOMIMUS
DROM-ee-see-oh-MIME-us

DRYOSAURUS
DRY-oh-SAWR-us

E

EDMONTOSAURUS
ed-MON-toe-SAWR-us

EINIOSAURUS
ay-nee-oh-SAWR-us

ELAPHROSAURUS
el-AFF-roh-SAWR-us

ELOPTERYX
ee-LOP-ter-iks

EODROMAEUS
EE-oh-DRO-may-us

EORAPTOR
EE-oh-RAP-tor

ERLIKOSAURUS
er-LIK-oh-SAWR-us

EUOPLOCEPHALUS
ewe-oh-plo-sef-ah-luss

EUSTREPTOSPONDYLUS
ewe-STREP-toe-SPON-die-luss

F

FABROSAURUS
fab-roh-SAWR-us

FALCARIUS
FAL-ca-ree-us

FERGANOCEPHALE
FER-gan-oh-sef-ah-ley

FERRISAURUS
FAIR-y-SAWR-us

FOSTEROVENATOR
FOS-teh-ro-VEN-ah-tor

FOSTORIA
FOS-tawr-ee-uh

FRUITADENS
FROO-ta-denz

FUKUIRAPTOR
FOO-KOO-ee-RAP-tor

FULGUROTHERIUM
FULL-gur-oh-THEER-ee-um

FUTALOGNKOSAURUS
FOO-ta-logn-koh-SAWR-us

G

GALLIMIMUS
gal-lee-MEEM-us

GASOSAURUS
GAS-oh-SAWR-us

GASPARINISAURA
GAS-pah-renn-ee-SAWR-us

GEMINIRAPTOR
GEM-y-ny-rap-tor

GIGANOTOSAURUS
gi-GAN-oh-toe-SAWR-us

GIGANTORAPTOR
gi-GAN-toe-RAP-tor

GIGANTSPINOSAURUS
gi-GANT-spy-noh-SAWR-us

GILMOREOSAURUS
GIL-more-oh-SAWR-us

GOBISAURUS
GOH-bee-SAWR-us

GOBIVENATOR
GOH-bee-VEN-ah-tor

GORGOSAURUS
GOR-goh-SAWR-us

GOYOCEPHALE
GOY-oh-sef-ah-ley

GRYPOSAURUS
GRIP-oh-SAWR-us

GUANLONG
gwan-long

H

HAGRYPHUS
HA-grif-us

HALSZKARAPTOR
HAL-z-ka-RAP-tor

HANSSUESIA
HANZ-su-es-ee-ah

HAPLOCHEIRUS
HAP-low-ky-rus

HARPYMIMUS
HAR-pi-MIME-us

HERRERASAURUS
herr-ERE-rah-SAWR-us

HESPERONYCHUS
hes-PARE-oh-NIKE-us

HESPEROSAURUS
hes-PARE-oh-SAWR-us

HETERODONTOSAURUS
HET-er-oh-DONT-oh-SAWR-us

HEXINLUSAURUS
HEY-zin-loo-SAWR-us

HEYUANNIA
hey-YOU-ARN-ee-ah

HOMALOCEPHALE
hom-ah-low-SEF-ah-ley

HUAXIAGNATHUS
hwax-EE-ag-NATH-us

HUAYANGOSAURUS
hway-YANG-oh-SAWR-us

HUEHUECANAUHTLUS
hway-hway-CAN-auh-tlus

HUNGAROSAURUS
HUN-gar-oh-SAWR-us

HYPACROSAURUS
HIE-pak-roh-SAWR-us

HYPSELOSAURUS
HIP-sel-oh-SAWR-us

HYPSELOSPINUS
HIP-sel-oh-spyn-us

HYPSILOPHODON
HIP-sil-OH-foh-don

I

ICHTHYOVENATOR
IK-thee-oh-VEN-ah-tor

IGUANACOLOSSUS
IG-wa-nah-co-LOSS-us

IGUANODON
IG-wa-noh-don

INCISIVOSAURUS
IN-si-see-vo-SAWR-us

IRRITATOR
IRR-eh-TAY-tor

J

JEHOLOSAURUS
JEH-hoe-lo-SAWR-us

JIANGXISAURUS
JEE-ang-ze-SAWR-us

JURAVENATOR
jur-AH-ve-NAY-tor

K

KAMUYSAURUS
KAH-mu-SAWR-us

KENTROSAURUS
KEN-troh-SAWR-us

KILESKUS
KY-les-kuss

KOL
KOL

KOSMOCERATOPS
KOS-mo-SER-ah-tops

KRITOSAURUS
KRI-toh-SORE-us

KULINDADROMEUS
KU-lin-dah-DRO-me-us

L

LAJASVENATOR
LA-jaw-VEN-ah-tor

LAMBEOSAURUS
lam-BEE-oh-SAWR-us

LANZHOUSAURUS
lan-ZOO-sawr-us

LAQUINTASAURA
LA-kwin-ta-SAWR-ah

LEAELLYNASAURA
lee-ELL-IN-a-SAWR-ah

LESOTHOSAURUS
le-SO-toe-SAWR-us

LIMUSAURUS
LIM-oo-SAWR-us

LINHENYKUS
LIN-hen-ei-kus

LOPHOSTROPHEUS
LOF-oh-stro-fee-us

LUANCHUANRAPTOR
LOO-an-chu-an-RAP-tor

LYCORHINUS
LIKE-oh-RINE-us

LYTHRONAX
LY-thron-ax

M

MAGNOSAURUS
MAG-no-SAWR-us

MAHAKALA
MA-HA-kah-la

MAIASAURA
my-ah-SAWR-ah

MAJUNGASAURUS
mah-JOON-gah-SAWR-us

MALAWISAURUS
mah-LAH-wee-SAWR-us

MAMENCHISAURUS
mah-MEN-chi-SAWR-us

MANIDENS
MAN-e-dens

MANTELLISAURUS
man-TEL-le-SAWR-us

MAPUSAURUS
ma-puh-SAWR-us

MASIAKASAURUS
mah-SHEE-ah-kah-SAWR-us

MEDUSACERATOPS
me-DEW-sa-SER-ah-tops

MEGALOSAURUS
MEG-ah-low-SAWR-us

MEI LONG
may-long

MICROPACHYCEPHALOSAURUS
MI-krow-PAK-ee-SEF-ah-loh-SAWR-us

MICRORAPTOR
MI-krow-RAP-tor

MICROVENATOR
MI-krow-ven-AY-tor

MINMI
min-mee

MIRAGAIA
MEE-rah-guy-ah

MIRISCHIA
mih-RISS-ke-ah

MOCHLODON
MOCK-loe-don

MONOCLONIUS
MON-oh-klo-nee-us

MONOLOPHOSAURUS
MON-oh-LOF-oh-SAWR-us

MONONYKUS
MON-oh-NEE-kus

MOROS
moe-ross

MUTTABURRASAURUS
MUT-ah-BUHR-a-SAWR-us

N

NANKANGIA
NAN-KANG-e-ah

NANUQSAURUS
NAH-nuk-SAWR-us

NANYANGOSAURUS
nan-yang-oh-SAWR-us

NASUTOCERATOPS
na-SU-to-SER-ah-tops

NEMEGTOSAURUS
ne-MEG-toe-SAWR-us

NIGERSAURUS
nee-gere-SAWR-us

NODOCEPHALOSAURUS
no-doe-SEF-ah-low-SAWR-us

NODOSAURUS
no-doh-SAWR-us

NOMINGIA
NOH-ming-ee-uh

NOTHRONYCHUS
NOH-thron-ike-us

NQWEBASAURUS
en-qweb-ah-SAWR-us

NYASASAURUS
Nye-ass-a-SAWR-us

O

OLOROTITAN
oh-LO-ro-TI-tan

OMEISAURUS
oh-mee-SAWR-us

ORNITHOLESTES
or-NITH-oh-LES-teez

ORNITHOMIMUS
or-NITH-oh-MEEM-us

ORODROMEUS
OR-oh-DRO-mee-us

ORYCTODROMEUS
OH-rik-toe-DRO-mee-us

OSTAFRIKASAURUS
OST-af-ree-kah-SAWR-us

OTHNIELOSAURUS
OTH-ni-EE-lo-SAWR-us

OURANOSAURUS
oo-RAHN-oh-SAWR-us

OVIRAPTOR
OH-vee-RAP-tor

OXALAIA
ox-AH-lie-ah

P

PACHYCEPHALOSAURUS
pack-ee-SEF-ah-loh-SAWR-us

PACHYRHINOSAURUS
pack-ee-RINE-oh-SAWR-us

PARALITITAN
pa-ra-li-TIE-tan

PARANTHODON
pa-RAN-thoe-don

PARASAUROLOPHUS
pa-ra-SAWR-ol-off-us

PARKSOSAURUS
PARKS-oh-SAWR-us

PARVICURSOR
PAR-ve-kur-sor

PATAGONYKUS
pat-ah-GONE-ei-kus

PATAGOTITAN
pat-ah-GOH-TIE-tan

PEDOPENNA
PE-do-PEN-nah

PENTACERATOPS
pen-tah-SER-ah-tops

PHILOVENATOR
FIE-loh-ven-AH-tor

PINACOSAURUS
PIN-ah-koh-SAWR-us

PISANOSAURUS
pee-ZAHN-oh-SAWR-us

PLATEOSAURUS
PLAT-ee-oh-SAWR-us

PNEUMATORAPTOR
new-MAT-oh-RAP-tor

PRENOCEPHALE
PREE-no-sef-ah-ley

PROA
proh-ah

PROCERATOSAURUS
proh-se-RAT-oh-SAWR-us

PROTOCERATOPS
pro-toe-SER-ah-tops

PSITTACOSAURUS
SIT-ak-oh-SAWR-us

Q

QANTASSAURUS
KWAN-ta-SAWR-us

QIAOWANLONG
ZHOW-wan-long

R

RAPETOSAURUS
rap-eh-to-SAWR-us

RATIVATES
RAT-i-VATE-eez

RHABDODON
RAB-doh-don

RHOETOSAURUS
reet-oh-SAWR-us

RINCHENIA
rin-CHEN-ee-ah

RUYANGOSAURUS
roo-YANG-go-SAWR-us

S

SAHALIYANIA
sa-HARL-ee-yan-ee-ah

SALTASAURUS
SALT-ah-SAWR-us

SALTOPUS
SALT-oh-pus

SALTRIOVENATOR
SAL-tre-oh-ven-AH-tor

SAUROLOPHUS
SAWR-oh-LOAF-us

SAUROPOSEIDON
SAWR-oh-POE-sy-din

SAURORNITHOIDES
SAWR-or-nith-OY-deez